THE IMPACTS OF RACISM AND BIAS ON BLACK PEOPLE PURSUING CAREERS IN SCIENCE, ENGINEERING, AND MEDICINE

PROCEEDINGS OF A WORKSHOP

Cato T. Laurencin, *Editor*

Cedric M. Bright and Camara P. Jones, *Rapporteurs*

Roundtable on Black Men and Black Women
in Science, Engineering, and Medicine

Policy and Global Affairs

Health and Medicine Division

The National Academies of
SCIENCES · ENGINEERING · MEDICINE

THE NATIONAL ACADEMIES PRESS
Washington, DC
www.nap.edu

THE NATIONAL ACADEMIES PRESS 500 Fifth Street, NW Washington, DC 20001

This activity was supported by contracts between the National Academy of Sciences and the Aetna Foundation. Any opinions, findings, conclusions, or recommendations expressed in this publication do not necessarily reflect the views of any organization or agency that provided support for the project.

International Standard Book Number-13: 978-0-309-67954-1
International Standard Book Number-10: 0-309-67954-0
Digital Object Identifier: https://doi.org/10.17226/25849

Additional copies of this publication are available from the National Academies Press, 500 Fifth Street, NW, Keck 360, Washington, DC 20001; (800) 624-6242 or (202) 334-3313; http://www.nap.edu.

Copyright 2020 by the National Academy of Sciences. All rights reserved.

Printed in the United States of America

Suggested citation: National Academies of Sciences, Engineering, and Medicine. 2020. *The Impacts of Racism and Bias on Black People Pursuing Careers in Science, Engineering, and Medicine.* Washington, DC: The National Academies Press. https://doi.org/10.17226/25849.

The National Academies of
SCIENCES · ENGINEERING · MEDICINE

The **National Academy of Sciences** was established in 1863 by an Act of Congress, signed by President Lincoln, as a private, nongovernmental institution to advise the nation on issues related to science and technology. Members are elected by their peers for outstanding contributions to research. Dr. Marcia McNutt is president.

The **National Academy of Engineering** was established in 1964 under the charter of the National Academy of Sciences to bring the practices of engineering to advising the nation. Members are elected by their peers for extraordinary contributions to engineering. Dr. John L. Anderson is president.

The **National Academy of Medicine** (formerly the Institute of Medicine) was established in 1970 under the charter of the National Academy of Sciences to advise the nation on medical and health issues. Members are elected by their peers for distinguished contributions to medicine and health. Dr. Victor J. Dzau is president.

The three Academies work together as the **National Academies of Sciences, Engineering, and Medicine** to provide independent, objective analysis and advice to the nation and conduct other activities to solve complex problems and inform public policy decisions. The National Academies also encourage education and research, recognize outstanding contributions to knowledge, and increase public understanding in matters of science, engineering, and medicine.

Learn more about the National Academies of Sciences, Engineering, and Medicine at **www.nationalacademies.org**.

The National Academies of
SCIENCES · ENGINEERING · MEDICINE

Consensus Study Reports published by the National Academies of Sciences, Engineering, and Medicine document the evidence-based consensus on the study's statement of task by an authoring committee of experts. Reports typically include findings, conclusions, and recommendations based on information gathered by the committee and the committee's deliberations. Each report has been subjected to a rigorous and independent peer-review process and it represents the position of the National Academies on the statement of task.

Proceedings published by the National Academies of Sciences, Engineering, and Medicine chronicle the presentations and discussions at a workshop, symposium, or other event convened by the National Academies. The statements and opinions contained in proceedings are those of the participants and are not endorsed by other participants, the planning committee, or the National Academies.

For information about other products and activities of the National Academies, please visit www.nationalacademies.org/about/whatwedo.

RACISM AND BIAS ACTION GROUP PLANNING COMMITTEE

CEDRIC BRIGHT (*Co-chair*), East Carolina University
CAMARA P. JONES (*Co-chair*), Morehouse School of Medicine
ANDRE L. CHURCHWELL, Vanderbilt University School of Medicine
CLYDE W. YANCY (NAM), Northwestern University

ROUNDTABLE ON BLACK MEN AND BLACK WOMEN IN SCIENCE, ENGINEERING, AND MEDICINE

CATO T. LAURENCIN (NAE/NAM)(*Chair*), University of Connecticut Health Center
OLUJIMI AJIJOLA, UCLA Medical Center
GILDA A. BARABINO (NAE), The City College of New York
CHARLES R. BRIDGES, JR., Janssen Research & Development, LLC
CEDRIC BRIGHT, East Carolina University
L.D. BRITT (NAM), Eastern Virginia Medical School
ANDRE L. CHURCHWELL, Vanderbilt University School of Medicine
THEODORE CORBIN, Drexel University
GEORGE Q. DALEY (NAM), Harvard Medical School
WAYNE FREDERICK, Howard University
PAULA T. HAMMOND (NAS/NAE/NAM), Massachusetts Institute of Technology
EVELYNN M. HAMMONDS (NAM), Harvard University
LYNNE M. HOLDEN, Montefiore Medical Center
CAMARA P. JONES, Morehouse School of Medicine
CORA BAGLEY MARRETT, University of Wisconsin-Madison
VALERIE MONTGOMERY RICE (NAM), Morehouse School of Medicine
RANDALL C. MORGAN, JR., W. Montague Cobb/National Medical Association
ELIZABETH O. OFILI (NAM), Morehouse School of Medicine
VIVIAN W. PINN (NAM), Senior Scientist Emerita, FIC, National Institutes of Health (Retired)
JOAN Y. REEDE (NAM), Harvard Medical School
LOUIS W. SULLIVAN (NAM), Morehouse School of Medicine
CLYDE W. YANCY (NAM), Northwestern University
MARK ALEXANDER (Ex Officio Member), 100 Black Men of America, Inc.
KIMBERLY BRYANT (Ex Officio Member), Black Girls CODE
GARTH N. GRAHAM (Ex Officio Member), Aetna Foundation
IAN HENRY (Ex Officio Member), Procter and Gamble Company
ORLANDO KIRTON (Ex Officio Member), Society of Black Academic Surgeons

JOHN R. LUMPKIN (NAM)(Ex Officio Member), Blue Cross Blue Shield of North Carolina Foundation
SHIRLEY MALCOM (Ex Officio Member), American Association for the Advancement of Science
ALFRED MAYS (Ex Officio Member), Burroughs Wellcome Fund
LAMONT R. TERRELL (Ex Officio Member), GlaxoSmithKline
HANNAH VALANTINE (Ex Officio Member), Stanford University

Project Staff

REGINALD HAYES, Program Officer, Board on Higher Education and Workforce
TOM ARRISON, Program Director, Policy and Global Affairs
PAULA W. WHITACRE, Consultant Writer

Preface

It is an honor for me to serve as chair of the National Academies Roundtable on Black Men and Black Women in Science, Engineering, and Medicine, as well as serve as the editor of the first in a series of proceedings publications from the Roundtable. Our work began in 2015 when leaders of the W. Montague Cobb/National Medical Association Health Institute and I recognized the growing absence of Black men in medical schools. In fact, levels of Black men entering medical school reached an historic low in the 2015 and 2016 years. Starting in 2016, and with financial support from important partners such as the Aetna Foundation, the Burroughs Wellcome Fund, the Robert Wood Johnson Foundation, and the Connecticut Black and Puerto Rican Legislative Caucus, we began planning a National Academies workshop on issues surrounding the absence of Black men in medicine. The joint workshop entitled "The Growing Absence of Black Men in Medicine and Science" took place in 2017. It was historic, in that to my knowledge it was the first National Academies activity specifically focused on issues involving Black people. The proceedings is entitled *An American Crisis: The Growing Absence of Black Men in Medicine and Science*. It was released in May of 2018, and corresponded to a briefing on the subject of Black men and medicine with the Congressional Black Caucus in Washington, DC. Many of the ideas that emerged from the workshop have been embraced by academia, industry, and philanthropy. More needs to be done.

Our next steps have involved the development of a more permanent presence in the National Academies to discuss issues surrounding Black men and Black women in science, engineering, and medicine. With support from our partners above, along with the Johnson and Johnson Company, the W.K. Kellogg Foundation, and the University of Pittsburgh—and with the continued leadership and commitment from Dr. Victor Dzau—president of the National Academy of Medicine, the Roundtable on Black Men and Black Women in Science, Engineering, and Medicine was launched late in 2018 as a joint activity of Policy and Global Affairs and the Health and Medicine Division. I am grateful to the steering committee members for the Roundtable: Drs. L.D. Britt, Cedric M. Bright, George Q. Daley, Randall C. Morgan Jr., Elizabeth Ofili, Vivian Pinn, and Louis Sullivan.

Our first formal meeting of the Roundtable took place in December 2019. It was decided that our first workshop should examine issues around racism and bias. Pernicious and pervasive, racism and bias in many ways serves as a backdrop to issues surrounding Black men and women in science, engineering, and medicine. I am grateful to the co-chairs of the workshop, Dr. Cedric Bright and Dr. Camara Jones. They expressly volunteered to take on our first Roundtable workshop, just 4 months after our first meeting and in the midst of the COVID-19 pandemic in April 2020. The workshop was held a month before the death of George Floyd. The ideas and concepts emerging from the workshop are especially important in these times.

> Cato T. Laurencin, M.D., Ph.D. (NAM/NAE)
> University Professor, The University of Connecticut
> Chair, Roundtable on Black Men and Black Women in Science, Engineering, and Medicine

Acknowledgments

This Proceedings of a Workshop was prepared by the workshop editor and rapporteurs as a factual summary of what was presented and discussed at the workshop. The planning committee's role was limited to planning and convening the workshop. The statements made are those of the editor and rapporteurs and do not necessarily represent positions of the workshop participants as a whole, the planning committee, or the National Academies of Sciences, Engineering, and Medicine. We wish to extend sincere thanks to all the members of the planning committee for their contributions in scoping, developing, and carrying out this project.

This proceedings has been reviewed in draft form by individuals chosen for their diverse perspectives and technical expertise, in accordance with procedures approved by the National Academies of Sciences, Engineering, and Medicine. The purpose of this independent review is to provide candid and critical comments to assist the institution in making its published report as sound as possible and ensure the document meets institutional standards for quality and objectivity. The review comments and draft manuscript remain confidential to protect the integrity of the process. We wish to thank the following individuals for their review of this proceedings: **Andre Churchwell**, Vanderbilt University; **Theodore Corbin**, Drexel University; **Samuel Mukasa**, University of Minnesota; and **Daryl Chubin**, Independent Consultant. Although the reviewers listed above have provided many constructive comments and suggestions, they were not asked to endorse the content of the proceedings, nor did they see the final draft before its release.

The review of this proceedings was overseen by **Maxine Hayes**, University of Washington (retired). Appointed by the National Academies, she was responsible for making certain that an independent examination of this proceedings was carried out in accordance with institutional procedures and that all review comments were carefully considered. Responsibility for the final content of this proceedings rests entirely with the rapporteurs and the institution.

Cedric M. Bright and Camara P. Jones
Co-chairs, Racism and Bias Action
Group Planning Committee

Contents

1 INTRODUCTION 1
   Opening Remarks, 3
   Organization of This Proceedings, 6

2 KEYNOTE ADDRESS: NEW ASPECTS OF RACISM 7
   Naming Racism Via Two Allegories, 8
   Identifying Action, 12
   Health Equity, 12
   References, 13

3 NOTES ON HISTORY, MYTH, AND RACE IN
   U.S. MEDICAL PRACTICE, 1619–2020 15
   African Americans as Providers of Medicine, 16
   Involvement of People of African Descent in Public Health, 19
   Health Care as a Right in the United States, 20
   Discussion, 20
   References, 21

4 SEGREGATION IN HOUSING AND EDUCATION 23
   The Myth of Residential Segregation as de facto Segregation, 24
   Government Policies to Create Residential Segregation, 26
   A Call for a New Civil Rights Movement, 28

Discussion, 28
References, 29

5 ATTACKS ON DIVERSITY, EQUITY, AND INCLUSION
IN EDUCATION   31
Historical Perspective, 32
Medical School Enrollment, 33
Graduate School Enrollment in Science and Engineering, 36
Medical School Faculty Representation, 36
Manifestations of Structural and Institutional Racism, 39
Discussion, 40
References, 42

6 RACISM IN RELATION TO BLACK MEN AND WOMEN
IN SCIENCE, ENGINEERING, AND MEDICINE   45
Factors Underlying the COVID-19 Mortality Gap, 46
Systemic Impacts and Need for Change, 49
Discussion, 51
References, 53

7 CONCLUDING REMARKS   55

APPENDIXES

A Workshop Agenda   57
B Biographical Sketches of Roundtable Members and
  Workshop Presenters   61

1

Introduction

Despite the changing demographics of the nation and a growing appreciation for diversity and inclusion as drivers of excellence in science, engineering, and medicine, Black Americans are severely underrepresented in these fields. For example, Black Americans constitute 13 percent of the U.S. population, but make up less than 7 percent of medical students and less than 3 percent of practicing physicians (NASEM, 2018). In the science and engineering disciplines, Black Americans represent 4.8 percent of employed professionals (National Science Foundation, 2015). As documented in several key reports, including *Altering the Course: Black Males in Medicine* (Association of American Medical Colleges, 2015) and *An American Crisis: The Growing Absence of Black Men in Medicine and Science* (NASEM, 2018), racism and bias are significant reasons for this disparity, with detrimental implications on individuals, health care organizations, and the nation as a whole.

In 2017, the Health and Medicine Division of the National Academies of Sciences, Engineering, and Medicine (the National Academies) and the Cobb Institute organized a national workshop that resulted in publication of *An American Crisis* (NASEM, 2018), mentioned above. That workshop served as a springboard for further discussion, with a number of articles and news stories on the subject appearing in print, and with acknowledgment that Black women face severe challenges as well. Among the discussion topics, it identified challenges encountered in the transition points to becoming successful physicians and scientists, including the impact of racism.

To address this and related issues, the Roundtable on Black Men and Black Women in Science, Engineering, and Medicine was launched at the National Academies in 2019. The Roundtable, comprised of 30 national leaders, has been charged to identify key levers, drivers, and disruptors in government, industry, health care, and higher education where actions can have the most impact on increasing the participation of Black men and Black women in these fields. The goals of the Roundtable are to:

- Compile and discuss quantitative and qualitative data relevant to the representation and experiences of Black men and Black women in science, engineering, and medicine;
- Convene a broad array of stakeholders representing higher education, industry, health care, government, private foundations, and professional societies;
- Highlight promising practices for increasing the representation, retention, and inclusiveness of Black men and Black women in science, engineering, and medicine; and
- Advance discussions that can lead to increasing systemic change.

The Roundtable focused its first public workshop on the role of racism and bias in the participation of Black men and Black women in science, engineering, and medicine (see Box 1-1 for the workshop Statement of Task). The workshop, entitled "The Impacts of Racism and Bias on Black People Pursuing Careers in Science, Engineering, and Medicine," took place on April 16, 2020.[1]

The workshop and this proceedings are intended to be an initial exploration of the context for the Roundtable's work; to surface key issues and questions that the Roundtable should address in its initial phase; and to reach key stakeholders and constituents. The workshop was not meant to be a comprehensive survey of the history and scholarship relevant to the complex topics of racism, bias, and impacts on Black men and Black women in science, engineering, and medicine.

This proceedings provides a record of the workshop discussions. The discussions covered racism (understood as conscious prejudice and antagonism), bias (which can operate unconsciously), and related barriers faced by Black men and Black women pursuing scientific, engineering, and medical

---

[1] Throughout this proceedings, the terms used for racial and ethnic groups are those used by each presenter.

INTRODUCTION 3

> **BOX 1-1**
> **Workshop Statement of Task**
>
> An ad hoc committee will organize a public workshop to examine the role of racism and bias in the decline of Black students in science, engineering, and medicine. The workshop will explore the historical trends of the enrollment of Black students in medical and engineering schools and the sciences, discussing the impact of the Flexner Report, the Bakke and Fisher decisions, other court challenges, and the U.S. Supreme Court decisions regarding the use of race in admissions. Some of the questions participants may consider are: What are the historical trends of Black faculty representation in science, engineering, and medicine? How does training on implicit and explicit bias mitigate the impacts of bias on Black students? Does explicit bias training combat the effects of racism and, if not, what else is needed? A brief rapporteur-authored workshop brief will be published.

careers. Several of the speakers addressing historical and statistical topics primarily focused on trends and developments in medicine and medical careers. This is partly a reflection of the fact that many of the Roundtable's inaugural membership cohort are medical doctors. These talks are summarized in Chapters 2 and 5. Chapter 5 also includes a discussion of trends and developments in graduate science and engineering fields. As the Roundtable continues to explore these topics in the future, it will focus on identifying and bringing forward experts and data sources that shed additional light on the Black men and Black women in science and engineering, as well as medicine.

OPENING REMARKS

Welcome and Background from the Roundtable Chair

Roundtable chair Cato T. Laurencin, M.D., Ph.D., University Professor and CEO of the Connecticut Convergence Institute for Translation in Regenerative Engineering at The University of Connecticut,[2] opened the workshop and emphasized that the "urgency of now" called for holding it vir-

---

[2] For full biographies of Dr. Laurencin and other presenters, see Appendix B.

tually, rather than postponing for a point after resolution of the COVID-19 pandemic would allow for travel. Dr. Laurencin explained that the Roundtable established six Action Groups to foster information-sharing and development of an evidence-based approach; engage with key stakeholders and broader communities of scientists, clinicians, engineers, and administrators; and design and conduct workshops, write papers and publications, and conduct activities for meaningful change. Each action group plans to hold a workshop and lead a data-gathering effort to coincide with the Roundtable's twice-yearly meetings. The current workshop, organized by the Racism and Bias Action Group, represents the first such effort, with future workshops organized by the five other groups: Public Advocacy; Mentorship and Advising; Psychological Factors; PreK to Graduate Education; and Financing.

Remarks from the President of the National Academy of Medicine

Victor Dzau, M.D., president of the National Academy of Medicine, welcomed participants and placed the workshop in the context of the COVID-19 crisis. He said the disproportionate impact of the disease on majority Black versus majority white counties is partly explained by differences in underlying health conditions and access to care. He noted:

> Many fewer primary care physicians and health care providers are from underrepresented racial and ethnic backgrounds. A diverse workforce has been identified as an important component of a quality and competent health care system. As Dr. Louis Sullivan [president emeritus of Morehouse School of Medicine and former secretary of the U.S. Department of Health and Human Services] has stated, "Failure to reverse these trends places the health of at least one-third of the nation's citizens at risk." And I think that's what we are seeing.

A shortage of Black men and Black women also persists in science and engineering, Dr. Dzau noted, which threatens the quality of the scientific enterprise and hampers progress for all. Lack of role models, lack of mentors, and socioeconomic factors are often mentioned to explain the causes of the shortage. But, he said:

> We have to recognize that bias and racism are the fundamental causes for the lack of Black men and women pursuing careers in

science, engineering, and medicine and the increasing health disparities as being bias and racism.... Discrimination, prejudice, and unconscious and conscious bias create exclusionary environments that prevent Black men and women from entering the pipeline and pursuing careers in science, engineering, and medicine. It is critical that we recognize that persistent structural racism and stereotyping still facing African American males and females is a significant problem, and unless we recognize it and start addressing this, we will not be able to address the underrepresentation of Black men and women in our fields.

Further Remarks

Vaughn Turekian, Ph.D., executive director of Policy and Global Affairs at the National Academies, thanked the Roundtable for leadership in pushing issues related to diversity and racism within the National Academies and, more broadly, in the nation and the world. "What we are seeing in this current crisis of COVID-19 is amplifying the very issues that we were raising before," he said. He expressed hope that the group can "pull a community and society to a better place to demonstrate the real positive outcomes by bringing people together."

The co-chairs of the Racism and Bias Action Group also spoke briefly at the start of the workshop to place the agenda in the current context. Camara Phyllis Jones, M.D., M.P.H., Ph.D., 2019-2020 fellow at the Radcliffe Institute for Advanced Study at Harvard University, stated:

The same things we are considering today in terms of racism and bias on Black people pursuing careers in science, engineering, and medicine are the same things that are causing more Black people to die from COVID-19. It's important for us to be naming racism again and still ... [P]ositioning the National Academies to come out on this topic, to saying the word 'racism' is very important.

Action Group co-chair Cedric Bright, M.D., FACP, associate dean at the Brody School of Medicine at East Carolina University, acknowledged the discussions during the workshop would at times be uncomfortable but would take place in a "safe zone." He asked participants to be "introspective, circumspective, and engaged as part of the discussion."

ORGANIZATION OF THIS PROCEEDINGS

The remainder of this publication is organized to follow the agenda of the workshop. Chapter 2 highlights the keynote address by Dr. Camara Jones on some of the causes and manifestations of racism. A presentation by Harriet A. Washington summarizes relevant medical and public health history in Chapter 3. Chapter 4 highlights Richard Rothstein's summary of federal legislation that segregated housing and education, rebutting the idea that residential segregation occurred in a de facto fashion. An overview of diversity, equity, and inclusion in higher education, with a focus on medical school application and enrollment data, was presented by David Acosta, M.D., and appears in Chapter 5. Howard Ross discussed racism in relation to Black men and Black women in science, engineering, and medicine, in Chapter 6. Chapter 7 presents concluding remarks. The workshop agenda and biographical sketches of the speakers can be found in the Appendixes.

REFERENCES

Association of American Medical Colleges. 2015. *Altering the Course: Black Males in Medicine.* Washington, DC. https://store.aamc.org/altering-the-course-black-males-in-medicine.html.

National Academies of Sciences, Engineering, and Medicine (NASEM). 2018. *An American Crisis: The Growing Absence of Black Men in Medicine and Science: Proceedings of a Joint Workshop.* 2018. Washington, DC: The National Academies Press. https://www.nap.edu/catalog/25130/an-american-crisis-the-growing-absence-of-black-men-in.

National Science Foundation. 2015. Employed Black Scientists and Engineers, as a Percentage of Selected Occupations: 2015. https://www.nsf.gov/statistics/2017/nsf17310/digest/occupation/blacks.cfm.

2

Keynote Address: New Aspects of Racism

Highlights of the Presentation

- Racism is a system of structuring opportunity and assigning value based on the social interpretation of how someone looks (which is what we call "race") that unfairly disadvantages some individuals and communities; unfairly advantages other individuals and communities; and saps the strength of the whole society through the wasting of human resources.
- Three types of racism can be delineated: institutional racism, personally mediated racism, and internalized racism.
- Two allegories—"The Two-Sided Sign" and "The Gardener's Tale"—help create an environment from which to discuss issues around racism.
- Mechanisms that can perpetuate any or all of the three types of racism include those related to structures, policies, practices and norms, and values.
- Health equity is the assurance of the conditions for optimal health for all people and must be present to eliminate health disparities.
- Barriers to health equity include a narrow focus on the individual, ahistorical stance, the myth of meritocracy, the myth of a zero-sum game, limited future orientation, myth of American exceptionalism, and white supremacist ideology.

> - Solutions include seeking ways to share a common humanity and finding power in collective action.

Camara Phyllis Jones, M.D., Ph.D., in addition to serving as a co-chair of the Roundtable's Racism and Bias Action Group, was the workshop keynote speaker. Her presentation provided an entry point to understand racism, its impact to confer disadvantages; advantages to individuals and the whole society; and mechanisms through which it operates.

Dr. Jones began by pointing out that although the original title of her presentation was "new aspects of racism," she chose to entitle it "again and still," especially in the context of COVID-19:

> It is my observation that most people in this country are still in denial that racism exists, that it has profound effects on the health and well-being of the nation. Every now and then, as with Hurricane Katrina, as with the Flint water poisoning, Charleston massacre, or Charlottesville marches, the general population is jolted out of it complacency.... As this pandemic wanes in a year and half or 2 years, we cannot let the country slip back into racism denial.

Dr. Jones explained that the goal of her presentation was to equip participants to do three things: name racism in a way to invite conversation; address how racism limits the representation of Black men and Black women in science, engineering, and medicine; and present theoretical principles for organizing and strategizing to act.

NAMING RACISM VIA TWO ALLEGORIES

In discussions about racism, Dr. Jones drew on allegories, which she said often help in airing issues that people are hesitant to talk about. She shared two such allegories at the workshop.

The first allegory she calls the "two-sided sign" (Figure 2-1). Briefly, she described an experience as a student at Stanford Medical School when she sat in a restaurant with friends and noticed a sign facing inside the restaurant that read "Open." She realized this meant the other side, facing outside, read "Closed." In other words, she continued, while the diners inside could enjoy a meal, those outside could not gain entry; moreover,

FIGURE 2-1 A depiction of the allegory "Dual Reality: A Restaurant Saga."
SOURCE: Camara Jones, Workshop Presentation, April 13, 2020, taken from Jones, 2016.

those inside might not even realize the existence of the sign or that it served to keep out prospective customers.[1] "I know it's hard to see if you only see 'Open,'" she said. "Part of your privilege is not to have to know" whether the privilege relates to race, sex, nationality, or other causes. But, she continued, "Once you *do* know, you can act.... Our challenge is that once we have a hint about a two-sided sign, we cannot forget this knowledge going forward."

Dr. Jones defined racism as:

> A system of structuring opportunity and assigning value based on the social interpretation of how someone looks (which is what we call "race") that unfairly disadvantages some individuals and communities; unfairly advantages other individuals and communities; and saps the strength of the whole society through the wasting of human resources.[2]

---

[1] For a fuller description, see Jones (2016).
[2] See also Jones (2003).

Breaking down this definition, Dr. Jones stressed racism as a system of power, rather than an individual's character flaw or moral failing. Race, she continued, is not a biological or cultural construct but a social interpretation of how one looks. In her own case, she noted she would be considered three different races in three different settings: "Black" in the United States, "White" in Brazil, and "Colored" in South Africa. Furthermore, "if I were to stay in any of those settings, my health and educational outcomes would become that of the group to which I was assigned, even with the same genes and abilities in all those places."

In terms of the impacts, Dr. Jones noted that the reciprocal to unfair disadvantages to some people is unfair advantages to others. "That is the whole issue of unearned white privilege that we hardly ever talk about because it makes some people uncomfortable," she said. While she said she used to apologize for bringing up this point, "Now I acknowledge that discomfort. I say if you feel that discomfort, lean in. Because I have come to recognize that the edge of our comfort is a growing edge." Referring to the last part of her definition above, sapping the strength of the whole society can be seen in not investing in full public education and thus missing out on genius to better the world, whether to find a cure for COVID-19, land humans on Mars, or any number of advances.

In her writing and research (e.g., Jones, 2000), Dr. Jones identified three levels on which racism operates:

- *Institutionalized racism*, "the constellation of structures, policies, practices, norms, and values that taken together result in differential access to goods, services, and opportunities of society by race." Examples she gave include differential housing, education, and employment opportunities, which also have an impact on health.
- *Personally mediated racism*, "differential assumptions about the abilities, motives, and intents of others by race, and differential actions based on those assumptions." Examples include instances of prejudice and discrimination, she said, including police brutality, physician disrespect, or teacher devaluation. Whether intentional or unintentional, and whether acts of commission or omission, personally mediated racism can affect Black men and Black women from reaching their potential in science, engineering, and medicine, she observed.
- *Internalized racism*, "acceptance by members of the stigmatized 'race' of negative messages about their own abilities and intrinsic

worth." Examples include self-devaluation, resignation, and hopelessness, and internalizing the myth of white superiority, including what is referred to as "the white man's ice is colder syndrome."

Another allegory that Dr. Jones often shares is called "The Gardener's Tale."[3] She developed it after seeing the difference in her own garden of flowers planted in rich, fertile soil compared with those planted in poor, rocky soil. In her allegory, a gardener preferred red flowers over pink and provided more favorable soil (institutional racism) to the seeds that would grow red flowers. The gardener began preferring the red flowers since they looked more vigorous and perhaps cut the pink flowers since they were not thriving (akin to personally mediated racism). Finally, the pink flowers wished they, too, could be red (internalized racism). Solutions to this disparity could begin with telling the pink flowers that they, too, are beautiful, but that is not sufficient, Dr. Jones said. Telling the gardener not to pluck the pink flowers might be helpful, but, as Dr. Jones said, the underlying issue—the fertility of the original soil or, to extend the allegory, the institutionalized racism—must be addressed.

Leaving the allegory, the "gardeners" are those who "have the power to decide, power to act, and have control of resources," Dr. Jones said, and include government, corporations, foundations, the media, communities to the extent they have self-determination, and institutions such as the National Academies. It is dangerous, she added, when the gardeners (or those in power) are allied with one group only and not concerned with equity. She said two questions often arise when she is sharing this allegory: Why should the red flowers share their soil, and what if the current gardener is not the one who made the decision to plant the two types of flowers in different soil in the first place? Extending beyond the allegory, she said:

> That is why we must make the problem of the pink flowers urgent. That is why we have this Roundtable on Black men and Black women in science, engineering, and medicine. We are problematizing it; we are putting it on the agenda. We will not be successful in addressing that inequity unless we understand the differences in the quality of the soil. Secondly, we must make those flower boxes transparent, talking about the differences in the quality of the soil. Third, we must also make sure that people understand

---

[3] For the full tale, see Jones (2000).

that the pink flowers did not just launch themselves into that poor, rocky soil. So, we must talk about history, and we must talk about how the gardener's initial preferences for red over pink set up the whole situation.

IDENTIFYING ACTION

In a given situation, Dr. Jones stated, it is important to ask how racism is operating in order to identify the levers on which to act. "That's what this workshop is going to bring us to," she said, to see how racism is operating in terms of Black underrepresentation in medicine. Related to structures (which she said are the "who, what, when, and where of decision making"), potential targets of interventions might include racial residential segregation, the presence or absence of quality preschool programs, and diversity of medical school faculty. Related to policies (the "written how of decision making"), solutions might include public school funding or affirmative action policies. Practices and norms (the "unwritten how") might include use of Medical College Admission Tests (MCATs) as a first hurdle to entry into medical school because of the norm that standardized tests are important predictors of success, while an applicant's "distance traveled" is considered only marginally relevant to success. Finally, values (the "why of decision making") might include the claim that Blacks are less intelligent or hardworking. Dr. Jones later returned to these levers of action in concluding remarks (see Chapter 6).

HEALTH EQUITY

Health equity is the assurance of the conditions of optimal health for all people. Achieving it requires valuing all individuals and populations equally, recognizing and rectifying historical injustices, and providing resources according to need, Dr. Jones said. She noted that health disparities can only be eliminated when health equity is achieved.

She identified seven barriers to achieving health equity, namely: (1) Narrow focus on the individual, which limits a sense of interdependence and collective efficacy, and makes systems and structures seem invisible or irrelevant; (2) an ahistorical stance, which disconnects the present from the past and makes the current distribution of advantage and disadvantage seem happenstance; (3) the myth of meritocracy, with the belief that "if you work hard, you will make it"; (4) the myth of a zero-sum game, which fosters competition over cooperation and hinders efforts to grow the pie; (5) a limited future orientation, unlike in other cultures that focus on the impact of present actions on future generations; (6) the myth of American exceptionalism, which prevents learning from others and creates a sense of U.S. entitlement; and (7) white supremacist ideology, which presents a hierarchy in human valuation with "white" as the ideal and the norm. She commented that, in particular, "The first three barriers are part of why this country continues to stay in denial about the existence and profound impacts of racism on the well-being of the nation."

Dr. Jones has developed three principles for achieving health equity, including:

- Valuing all individuals and populations equally
- Recognizing and rectifying historical injustices
- Providing resources according to need.

Given the barriers and the need to put these principles into practice, Dr. Jones suggested a number of actions and strategies to workshop participants to increase the numbers of Black men and Black women in science, engineering, and medicine. As in the earlier analogy of the restaurant, she urged looking for evidence of two-sided signs, bursting through the bubble to experience common humanity, and becoming interested and joining in the stories of others. She further suggested taking notice of absences to "develop the skills to see who is not at the table, what is not on the agenda." She concluded by stressing the importance of revealing inaction in the face of need, and recognizing that action, especially collective action, is power.

REFERENCES

Jones, C. P. 2000. Levels of racism: A theoretic framework and a gardener's tale. *American Journal of Public Health*, 90(8), 1212–1215.

Jones, C. P. 2003. Confronting institutionalized racism. *Phylon*, 50(1-2), 7–22.

Jones, C. P. 2014. Systems of power, axes in inequity: Parallels, intersections, braiding the strands. *Medical Care,* 52(10 Supplement 3), S71–S75.

Jones, C. P. 2016. How understanding of racism can move public health to action: Allegory highlights reality of privilege. *The Nation's Health,* 46(1), 3.

3

Notes on History, Myth, and Race in U.S. Medical Practice, 1619–2020

Highlights of the Presentation

- History has presented falsehoods as "scientific truth," such as the claim that African Americans could not and did not practice medicine or engage in scientific research.
- Many men and women of African descent were skilled health care providers or innovative researchers, often not receiving credit for their contributions.
- The Flexner Report, published in 1910 with support from the American Medical Association, severely curtailed access to medical training for African Americans under the guise of modernization of the profession.
- Claims that African Americans have different or inferior traits have led to situations in which they have been used as subjects for inappropriate medical research.
- Conversely, people of African descent are often not provided cutting-edge medical discoveries, such as ZMapp to treat Ebola, under the grounds that they could not understand how to comply with treatment.
- African American medical ethicists with a firm grounding in history are needed to ensure perspective in analysis and research.

Harriet A. Washington, author of *Medical Apartheid: The Dark History of Medical Experimentation on Black Americans from Colonial Times to the Present* and other publications, drew on her research to discuss three main topics: the history of the involvement of people of African descent in medicine, the history of people of African descent in public heath, and the history of proposed health care as a right in the United States.

This history is important because it has a bearing on health today, she said. As quoted in *Unequal Treatment: Confronting Racial and Ethnic Disparities in Health Care* (Institute of Medicine, 2002, p. 6), "racial and ethnic disparities in health care occur in the context of broader historic and contemporary social and economic inequality and evidence of persistent racial and ethnic discrimination in many sectors of American life." As she described in her presentation:

- Historical patterns of approach to "Black diseases" subtly inform today's approach to racially disparate disease patterns;
- Abusive, stigmatizing research has fed iatrophobia (fear of doctors) by many African Americans;
- History must inform medical ethics;
- History lives in today's issues.

AFRICAN AMERICANS AS PROVIDERS OF MEDICINE

The history of African American involvement in science and medicine has perpetuated many inaccuracies, Ms. Washington noted. "Science and history are both privileged disciplines and are treated as the gospel truth," she said, "but they can be edited and curated to preserve a certain perspective." She summarized a number of falsehoods that were presented as science-based truth. For example, a 1913 article written by the editor of *Science* magazine (Cattell, 1913) stated, "There is not a single mulatto who has done credible scientific work." As another example, an organization called the American College of Ethnology developed "scientific racism" to justify slavery and profitable medical research, claiming that African Americans were a "hardy" different species than white humans. A number of imaginary diseases were ascribed to them, Ms. Washington continued, as well as a false scientific basis for traits such as low intelligence, poor judgement, poor parenting, and the inability to control "bestial sexual urges." The U.S. Census of 1840 was purported to reveal that free Blacks suffered 11 times the insanity rate of enslaved Blacks.

Despite the myth that they could not be trusted as doctors, African American women and men had a vast amount of medical knowledge. Women were known for their skills as midwives, herbalists, and healers, including by their white enslavers. In fact, the midwives often had better outcomes than white male obstetricians.

The medical profession imposed a number of barriers to African Americans practicing medicine. When a bad outcome did occur, Black health care providers often had to provide character witnesses to avoid being accused of poisoning a patient. In the 1800s, African American men could attend medical school but not earn a degree, and the American Medical Association (AMA) designated a "C" next to the names of African American physicians to denote "colored." In the 1850s, Harvard Medical School expelled its first three Black students because of pressure by white students. Into the 1960s, an overtly racial bar or subtler exclusion persisted at many medical schools as a way to limit the number of physicians and promote exclusivity.

As part of the impetus toward professionalism, the AMA sponsored what became known as the Flexner Report, published as a report to the Carnegie Foundation for the Advancement of Teaching in 1910. Ms. Washington refuted the argument made by some people that the AMA did not actually sponsor the report, noting that author Abraham Flexner had an office and staff at the AMA headquarters in Chicago and an AMA staff member accompanied him on his site visits to medical schools. According to Ms. Washington:

> The Flexner Report is sometimes referred to as a milestone in the modernization of medicine, but for African American practitioners, it was a disaster. It recommended the closing of every Black medical school except for two. Only Howard and Meharry were left. Even worse is the language of the report, which I do not think gets as much attention as it should. Abraham Flexner, although not a physician or with a Ph.D. of any kind, had strong opinions, certified by the AMA at that time, about who should become a doctor.

The views contained in the report, Ms. Washington continued, included that African Americans should not be allowed to practice except under the supervision of whites, should not be trained in medical specialties or research, and should have sharply curtailed duties. The report acknowledged the need for some limited provision of health care by African

Americans to African American patients, primarily to keep infectious disease from white communities. (See also Chapter 5 for a discussion of the ramifications of this report).

Although the AMA did not technically discriminate against African Americans after 1847, its feeder organizations on the local level did, and membership to the national organization could only take place through membership in a feeder organization. Although African American physicians protested, including picketing at AMA conventions, many were barred from residencies and hospitals. "When [African American] physicians were barred from hospitals, the effect on African American communities was disastrous," she said. "They could not follow their patients into hospitals. Many Black patients started to view Black physicians as inferior. Yet, white doctors often had no place for [Black patients]."

In 2008, Ms. Washington was part of an expert panel that investigated and wrote about the AMA's past record of discrimination. After their article, "African American Physicians and Organized Medicine, 1846-1968," appeared in *JAMA* (Baker et al., 2008), the AMA issued an apology. "It was important," she said. "It was accompanied by some action and a long overdue acknowledgment."

Despite being discouraged or barred from research, many African Americans did participate in groundbreaking research, often as uncredited contributors of teams lead by whites. "They were a shadow army of medical researchers who got no acknowledgment," said Ms. Washington, who is currently researching their lives, some of whom were collectively known as the "Garçons" (French for "boys"). "It's important for them to be better known, especially for African American young people, so they see that far from being an exotic or unattainable goal, it's something we have always succeeded in," she said. Twentieth-century examples include heart researchers Vivien Thomas and Hamilton Naki, chemist Percy Julien, and many others. Earlier examples include Onesimus, who provided Cotton Mather with knowledge about smallpox vaccinations, and James McCune Smith, who went to Scotland for his medical education and, among other accomplishments, disproved the claim from the 1840 U.S. Census about the mental health impacts of freedom for African Americans.

The belief that African Americans are "mentally inferior" or "inherently violent and unintelligent," referred to by Ms. Washington as "hereditarian follies," continues to lead to questioning of the qualifications of African American physicians. As recent examples, she noted several incidents on airplanes when a passenger was having a medical emergency and the flight attendants demanded the credentials of African American female physicians who volunteered to help, while not doing the same to white male physicians.

INVOLVEMENT OF PEOPLE OF AFRICAN DESCENT IN PUBLIC HEALTH

Erroneous beliefs about African Americans have also led to situations in which they become the subjects of inappropriate or harmful medical research. Lower intelligence was supposedly "verified" in seemingly painstaking but rigged research. In a few high-profile cases, Ms. Washington said, there has been an "illusion of Black complicity": that is, that African American physicians and caregivers were actively involved in the studies. One of the most well-known is the Tuskegee Syphilis Study conducted by the U.S. Public Health Service. Ms. Washington pointed out that Eunice Rivers, the African American woman who organized the participants, is often mentioned in descriptions of this project. In her investigation, Ms. Washington said:

> I found it interesting that almost anybody who has read the study can tell you who Eunice Rivers was, but no one can name the men who were the actual architects of the study, the scientists who actually carried it out and designed it and deflected the blame on her. When the initial reports were written up, the authors were all white. When people started impugning it [the study], suddenly the papers included her as one of the authors.

In addition, a widely circulated photograph of the study team included an African American man. Ms. Washington investigated to learn that the man had nothing to do with the study. Another frequently mentioned example is Solomon McBride, who conducted abusive research studies at Holmesburg Prison in Pennsylvania, but he was a prison official and not a physician. As an example of a more recent manifestation of the tendency to devalue African Americans in research, Ms. Washington pointed to myths around epidemic management. ZMapp, a treatment against Ebola, was given to a white U.S. physician rather than Africans with the disease, including Sierra Leone's chief virologist, Sheik Umar Khan, M.D., based on the beliefs that Africans are not intelligent enough to take the new drugs correctly or other misperceptions.

These cases have led Ms. Washington to conclude:

> One area where I see a need for far more African American participation is for African American medical ethicists to promulgate their analysis and research. I see too much research being weighed, valued, and put into effect with no regard to the populations on whom it is going to resonate....The law has changed to permit research without consent, and the people who are deciding if this is an appropriate or ethical thing to do are not the people who end up being the subjects of that research. We need someone speaking up for them. I very much hope we see more Black people going into medical ethics.

HEALTH CARE AS A RIGHT IN THE UNITED STATES

Mythologies related to medicine can also foment disparities. Historically, this is shown under slavery where fitness for work, not health, was the goal. The primary relationship, or dyad, was between the physician and the planter, rather than between the physician and the patient. Today, she said, the dyad is often between the physician and the state, and Black people are increasingly seen as research subjects or are underserved. Another tendency, she said, is medical dimorphism, or blaming the victim, rather than disparate treatment and environment, as the cause of contemporary "Black diseases"; that is, disease that disproportionately affect African Americans, such as diabetes and cardiovascular conditions.

DISCUSSION

Related to the point about African Americans as subjects of research, a question was raised about recent statements by a French scientist who wanted to experiment with COVID-19 vaccinations on Africans.[1] Ms. Washington replied, "What they voiced is not novel." She said that countries in the developing world with less oversight and regulation are becoming the "laboratory for the West."

Dr. Bright asked about the impact of the Flexner Report on medical school admissions, such as the use of standardized tests as an important

---

[1] For an account of this statement, see https://www.bbc.com/news/world-europe-52151722 [Accessed May 12, 2020].

criterion of admission. Ms. Washington noted that one goal of the report was to make medicine "exclusive," even if "we are not necessarily careful and precise about what we exclude." As an analogy, she said knowledge of German could be a prerequisite to exclude many candidates, yet is not one that predicts medical success. She suggested studies to show the correlation, if any, between test scores and performance in medical school and as a physician. "If the idea is to have enough doctors for the American people, we must be careful to make sure the tests that we give people are really going to be predictive of abilities as a clinician," she said. "If there is not a correlation, we should discard them [standardized tests] and find something that does."

In answer to a question about preparation for a career as a medical ethicist, Ms. Washington suggested fellowships for those who already have a medical background or need basic grounding, as well as master's or Ph.D. degrees in medical ethics for those who need more depth, such as to serve as expert witnesses. In all cases, "learning about the history of medicine is important," she said. "I am concerned that ethicists don't have a good grounding in history."

When a participant shared her concern that ventilators may not be made as available to Black patients suffering from COVID-19, Ms. Washington expanded the conversation to consider allocations of all scarce resources in a health care setting. "There are always criteria and policies that look neutral and may be intended to be neutral but they are not," she said. "Clinicians follow the rules, but in the rules are judgment calls," citing organ transplant policies as another example. Liver disease is a leading cause of death among African American men, she noted, but many are not receiving liver transplants because they are deemed not to have a social support system, which is one of the criteria in selecting who receives a new liver. Camara Jones, referring to the three principles of health equity she identified in her talk (see Chapter 2), said she has concluded the best way to allocate scarce resources is through a random lottery.

REFERENCES

Baker, R. B., H. A. Washington, O. Olakanmi, et al. 2008. African American physicians and organized medicine, 1846-1968: Origins of a racial divide. *JAMA*, 300(3), 306–313. doi: 10.1001/jama.300.3.306.

Cattell, J. M. 1913. *Science*, 39(1004), 5.

Davis, R. M. 2008. Commentary: Achieving racial harmony for the benefit of patients and communities. *JAMA*, 300(3), 323–325.

Institute of Medicine. 2002. *Unequal Treatment: Confronting Racial and Ethnic Disparities in Health Care.* Washington, DC: The National Academies Press.

Washington, H. A. 2006. *Medical Apartheid: The Dark History of Medical Experimentation on Black Americans from Colonial Times to the Present.* New York: Random House.

4

Segregation in Housing and Education

Highlights of the Presentation

- By the end of the 1960s, legal segregation was abolished but residential segregation existed and continues to exist in every metropolitan area in the United States.
- Rather than being considered "de facto" segregation resulting from private sector activity, federal policy established and has perpetuated residential segregation.
- The achievement gap in education can be explained by residential segregation because unequal social and economic conditions that impact academic performance are disproportionately present in segregated neighborhoods, which then feed into segregated schools.
- Federal housing policies included clauses that prohibited sale, resale, or rental of homes in post-World War II suburban developments to African Americans. Public housing beginning in the 1930s was also explicitly racially segregated.
- On average, African American households have 7 percent of the wealth of white households; wealth derived in large part from the equity built in these homes.
- A new civil rights movement is necessary to create the atmosphere in which policies to remedy residential segregation can be enacted to counter educational, health, and other consequences.

Richard Rothstein, Distinguished Fellow at the Economics Policy Institute, spoke on ways in which the past has determined race-based social inequality today in the United States. He focused his presentation on federal housing policies that have resulted in residential segregation, countering the widely held assertion that these housing patterns are "de facto" because they emerged from individual or market-based activity.

THE MYTH OF RESIDENTIAL SEGREGATION AS DE FACTO SEGREGATION

In the first part of the 20th century, the civil rights movement began by challenging segregation in law schools, Mr. Rothstein said, and then went on to challenging segregation in colleges and universities, leading to Brown v. Board of Education in 1954.[1] "The Brown decision inspired, motivated, and energized a growing civil rights movement," he said. "By the end of the 1960s, the movement had persuaded much of the country—not all of the country—that racial segregation was wrong, harmful to Blacks and whites, and was incompatible with our self-conception as a constitutional democracy." By the end of the 1960s, legal segregation was abolished in public accommodations, transportation, schools, employment, and other areas of life. However, Mr. Rothstein said:

> The civil rights movement disbanded and left the biggest segregation of all: which is that every metropolitan area in this country is residentially segregated.... What we've done, all of us, Blacks and whites, liberals and conservatives, Democrats and Republicans, northerners and southerners, is to create a rationalization to explain the failure.... We tell ourselves what happened by accident can only un-happen by accident. The name we give to this myth is de facto segregation.

The rationalization, he continued, is to say that residential segregation was not caused by the government, but rather by bigoted landlords or homeowners, by banks and real estate agents in the private economy, or because of personal preferences or difference in income level—terming it de facto segregation. "We regret it, but we don't think it is our responsibility to fix

---

[1] For the full case, see *Brown v. Board of Education*, 347 U.S. 483 (1954), https://www.loc.gov/item/usrep347483 [Accessed May 20, 2020].

it," he said, paraphrasing the belief that the government did not cause nor can it solve residential segregation.

According to Mr. Rothstein, residential segregation lies behind the achievement gap between Black and white students. He took issue with the premise of No Child Left Behind (NCLB), which he noted passed in 2001 with bipartisan support and was endorsed by the civil rights community.[2] NCLB, he said, ascribed the achievement gap to teachers' low expectations of low-income, particularly Black, students. Mr. Rothstein countered this premise by pointing out that children who live in more polluted neighborhoods, in poorly maintained buildings, and with other disadvantages may have health conditions that impact their learning. "What happens if almost all the kids in the school have these conditions, how can you expect that school to have the achievement level of a school where most children are well rested and healthy? It is impossible to have that expectation," he said. Mr. Rothstein said:

> Schools today are more segregated than at any time in the last 50 years in this country, and they are more segregated because the neighborhoods in which they are located are segregated. I came to the conclusion that neighborhood segregation is an educational problem and not just a housing problem.

The premise that residential segregation is a form of de facto segregation was cited in a 2007 U.S. Supreme Court decision about school-choice lottery systems in Louisville and Seattle.[3] The case involved a situation in which a Black child and a white child might be vying for one remaining slot in a school, and consideration for diversity might give the Black child the slot (according to Mr. Rothstein, a situation that rarely arises in reality). In his controlling opinion, Chief Justice John Roberts wrote that de facto neighborhood segregation caused the school segregation, and the U.S. Constitution prohibits taking explicit action to redress the situation.

---

[2] No Child Left Behind (NCLB) Act of 2001. Pub. L. No. 107-110.

[3] For background on the case, see Greenhouse, L. 2007. Justices limit the use of race in school plans for integration. *New York Times.* https://www.nytimes.com/2007/06/29/washington/29scotus.html [Accessed May 13, 2020].

GOVERNMENT POLICIES TO CREATE RESIDENTIAL SEGREGATION

Deliberate government actions, rather than de facto causes, have created neighborhood segregation, Mr. Rothstein said. Using Louisville as an example, he noted when a white homeowner purchased and re-sold a home to an African American family in the community of Shively in the 1950s when the family could not purchase a home otherwise. Mobs protected by the police firebombed the home, and the white homeowner was tried and convicted for sedition. Police protection of the mob, Mr. Rothstein pointed out, was a violation of the 14th Amendment to the U.S. Constitution. "This was not a unique situation," Mr. Rothstein continued. "In the 20th century, there were thousands of cases of police-protected mobs driving African Americans out of their homes that they had purchases or rented in white neighborhoods. Every one of them was a 14th Amendment violation that we have never remedied." Although bigotry, income differences, and some self-segregation play into housing patterns, he said, none of them is powerful enough to explain the segregation that continues today. "Inequality today is not merely the legacy of slavery—not to minimize that—it is the ongoing effects of Jim Crow policies enforced by the government that we still live with today."

Racially explicit federal policies have led to residential segregation and a wealth gap between Blacks and whites, Mr. Rothstein said. The most important, according to Mr. Rothstein, was the policy of the U.S. Federal Housing Administration (FHA) and the Veterans Administration (VA) to move the white working class from urban areas into single-family homes in the suburbs, such as Levittown, east of New York City; Lakewood and Panorama City near Los Angeles; and Westlake and Daly City near San Francisco, and many others. Developers could not have assembled the necessary capital to build such extensive developments on their own and needed public financing. The FHA and VA required developers to place a clause in the deed of homes prohibiting sale, resale, or rentals to African Americans. "This was not the action of rogue bureaucrats at the FHA or VA," he said. "It was written out in a federal policy manual distributed to appraisers all over the country." The manual further stated that federal financing guarantees would not be given to integrated developments or even to all-white developments located near an African American neighborhood.

The consequences of this policy can be shown in a gap in wealth decades later, Mr. Rothstein said. Returning white war veterans could buy a house for about $9,000 at the time, with no down payment and a mortgage that was less than the rent they were paying in public housing. As they built equity, they could send children to college, save for retirement and emergencies, and bequeath wealth to their children and grandchildren. "African Americans were prohibited by federal policy from participating in this wealth-generating exercise," Mr. Rothstein said. On average, according to Mr. Rothstein, African American family incomes are 60 percent of white incomes, but the wealth disparity is more disparate: African American households have, on average, 7 percent of the wealth of white households.

He said:

> That enormous gap between a 60 percent income ratio and a 7 percent wealth ratio is entirely attributed to unconstitutional federal housing policies that has never been remedied, that we have never taken as an obligation to remedy. We are clouded by this myth of de facto segregation. We have forgotten this history.

Other federal, state, and local government policies have also created residential segregation, Mr. Rothstein continued. As an example, he explained that public housing began during the New Deal when the private sector was not building residences. The Public Works Administration built housing for working families across the country, and that housing was segregated. In some cities, such as Cleveland, Atlanta, and Cambridge, Massachusetts, housing previously integrated neighborhoods was demolished and gave way to construction of larger, segregated building development. During World War II, when people moved to cities to work in war production factories, the government had to quickly build more housing, which was segregated. This is particularly noteworthy on the West Coast, he commented, which had previously seen less African American migration than other parts of the country.

After World War II, the housing shortage persisted. President Truman proposed an expansion of public housing to provide for returning veterans. Although the rents were not subsidized, conservatives in Congress opposed the government-provided housing because they considered it socialistic, Mr. Rothstein explained. They came up with a legislative strategy, known as a "poison pill," to insert an amendment in the bill that public housing had to be non-discriminatory, reasoning that the amendment would result in defeat of the overall bill. According to Mr. Rothstein, northern liberals voted against the amendment in order to preserve the chances for the larger public housing bill to pass. One result is that the federal government used the vote against the non-discrimination amendment as a justification to continue to segregate public housing explicitly for the next 15 years. In one respect, he said, the country is making a similar housing decision today. Faced with a continued shortage of housing that disproportionately affects African American and Hispanic families, federally subsidized affordable housing is built in low-income areas, reinforcing segregation, because it is "easier" than building these homes in high-opportunity neighborhoods where there might be community-organized opposition.

A CALL FOR A NEW CIVIL RIGHTS MOVEMENT

Mr. Rothstein summarized what he said are the consequences of these federal policies of segregation. They include the achievement gap created from concentrating children with low social and economic conditions in single schools; health disparities from living in areas that are more polluted and stressful than where whites live, on average; disproportionate incarceration of African American men who are concentrated in neighborhoods without jobs or transportation to get to jobs in other neighborhoods (not discussed above); the wealth gap described above; and political polarization. "How can we expect to create the common national identity that is necessary to preserve this democracy if so many African Americans and whites live so far from each other?" he asked.

The policies to remedy these consequences are well known to policy makers, Mr. Rothstein asserted. He called for a "new civil rights movement to create the atmosphere to make it uncomfortable to maintain the policies of segregation, just as the past civil rights movement made it uncomfortable to maintain other forms of segregation in the 20th century." He concluded, "with this much more accurate and passionate public discussion of racial inequality and the legacies of slavery and Jim Crow, I am hopeful we can develop a new civil rights movement that will redress residential segregation and take care, at least in part, of the consequences I have described."

DISCUSSION

Related to the achievement gap in education, a participant asked Mr. Rothstein about the effect of online learning resulting from COVID-19 school closures. He referred to an article he wrote on this topic (Rothstein, 2020). In addition to a gap in access to computers and high-speed Internet, there are disparities in how parents can help their children with home learning. "The current achievement gap between Black and white children is estimated to be about 2 years of schooling," he said. "I estimate that the coronavirus will add another half-year." He called for vastly increased resources to low-income schools to offset the effect of the coronavirus on the achievement gap.

Another participant commented that FHA and VA discrimination is not common knowledge and asked how to increase public understanding of this history. Mr. Rothstein said the way is to learn about the situation and tell others, as he did in his book *The Color of Law* (Rothstein, 2017). "It is not new history, it is not hidden," he said. "If we understand that segregation that we see everywhere in the country was created by government, unconstitutionally, with policy, we can understand that policy can do something about it."

A participant asked Mr. Rothstein's view of standardized tests that are used to gain entry to magnet schools. He urged an abolishment of standardized tests for admittance, noting that many colleges are making the SAT optional. "I hope that policy and understanding will trickle down to selective high schools," he said. "These tests are not an accurate measure of student ability to succeed."

Several questions related to disparities, including the achievement gap, that exist in middle-class African American areas were asked. Mr. Rothstein said African American middle-class households tend to live closer to lower-income neighborhoods than middle-class whites do and are more susceptible to the disadvantages in lower-income neighborhoods. He noted other factors have an impact on middle-class African Americans, including experiences of discrimination and the wealth gap described above.

Mr. Rothstein called for a political force to demand more funding for low-income schools. He clarified that educational funding does not wholly depend on property taxes. "It is not a technical question," he said. "We know how to redress inequality in housing and schools. We don't just need equal funding, we need much higher funding for children from lower social class backgrounds, who are predominantly African American and Hispanic. There is a need to create the political environment."

REFERENCES

Rothstein, R. 2017. *The Color of Law: A Forgotten History of How Our Government Segregated America*. New York: Liveright Publishing.

Rothstein, R. 2020. The coronavirus will explode achievement gaps in education. *Shelterforce: The Voice of Community Development.* https://shelterforce.org/2020/04/13/the-coronavirus-will-explode-achievement-gaps-in-education.

5

Attacks on Diversity, Equity, and Inclusion in Education

Highlights of the Presentation

- An understanding of history, including the Flexner Report, is important to move the needle forward in increasing African American representation in science, engineering, and medicine.
- Medical schools were not required to integrate until the 1960s, and, in 1968, 130 African Americans graduated from all U.S. medical schools.
- Applications and matriculation of African Americans in medical school and in graduate science and engineering programs have remained relatively flat over the past 30 to 40 years.
- Pipeline and other initiatives have resulted in slight increases, while anti-affirmative action laws have had a negative impact.
- Faculty numbers have also remained low, and minority faculty, particularly women, report significant levels of harassment, discrimination, and other negative actions.
- Data are needed to provide evidence about graduation rates, the students who take the MCAT but do not apply to medical school, and disciplines such as engineering that have particularly low rates of Black and African American enrollment.

David Acosta, M.D., FAAFP, chief diversity and inclusion officer at the Association of American Medical Colleges (AAMC), reviewed the historical trend of African American/Black students in medicine, engineering, and the sciences; analyzed the impact of anti-affirmative action laws and other barriers on diversifying the healthcare and science workforce; reviewed the historical trend of African American/Black faculty representation in medicine, engineering, and the sciences; and described the manifestations of structural and institutional racism in academic health science centers and the impact on medical students, residents, and faculty.[1]

HISTORICAL PERSPECTIVE

Reflecting on the earlier presentations, Dr. Acosta stressed the need to understand history in order to move forward. It is important to be proactive in telling this history. He noted many medical educators and diversity officers do not have in-depth knowledge about the Flexner Report and, more broadly, the long-term impact of the U.S. Supreme Court Plessy v. Ferguson case that provided the legal justification for segregation.[2] Abraham Flexner, he said, "essentially codified the two Americas in his plans for improving medical education for U.S. physicians." A two-page chapter of the report entitled "The Medical Education of the Negro" stated that this education was to promote "the limited education of the African American doctor as a service to his own race" and for the purpose of keeping whites from the spread of disease among African Americans. Flexner stipulated that medical schools had to be connected with universities with sufficient endowments and hospitals. This requirement was a severe challenge for most medical schools that taught African Americans at the time, which Flexner termed had good intentions but were "make believe" institutions. Five of the seven existing schools closed.

Universities, including medical schools, did not have to comply with desegregating their institutions until after Brown v. Board of Education in 1954 and the Civil Rights Act of 1964. In 1968, according to Dr. Acosta, 130 African Americans graduated from all U.S. medical schools. At that time, the American Medical Association (AMA) and AAMC joined to say that medical schools should have as a goal to expand the enrollment to a

---

[1] Dr. Acosta explained that his data use both "Black" and "African American" to account for students both born in and outside the United States.

[2] For more background, he suggested Steinecke and Terrell (2010).

level that permits all qualified applicants, although did not specifically refer to Black students. That same year, AAMC encouraged member medical schools to collect enrollment information by race and ethnicity, and, in 1969, established an Office of Minority Affairs. Until 1988, however, the office was embedded in the organization's Student Affairs Division and led by a non-physician. In 1988, AAMC created the Division of Minority Health Education and Prevention and hired its first vice president to lead diversity work, Herbert Nickens, M.D.

As Dr. Acosta transitioned to a discussion about application and enrollment figures, he commented, "Do not forget the history when you look at where we are today."

MEDICAL SCHOOL ENROLLMENT

Dr. Acosta first presented data related to medical school application and matriculation rates from 1980 to 2018.

Medical School Applications

In 1980, according to AAMC data, 2,507 Black or African American students applied to medical school, 4,344 applied in 2016, and 4,438 in 2018. He characterized this as a small increase numerically and percentage-wise, going from 7 percent of applicants to 8.2 percent in 2016 and 8.3 percent in 2018. He also called attention to the total number of applicants of all races at those different points in time, which increased from 36,083 in 1980 to 53,042 in 2016, with a slight dip to 52,277 in 2018 (Table 5-1).

Dr. Acosta cautioned the audience to look carefully at how data are compiled and reported. For instance, some AAMC reports show that Blacks and African Americans applicants represented 9.8 percent of the total in 2018–2019, rather than the 8.3 percent mentioned above. He attributed the larger number to how students of combined races or ethnicities are asked to fill out the American Medical College Application Service (AMCAS) application.

In addition to overall historical trends, Dr. Acosta said it is important to look at what he termed the influencers and detractors that affected application numbers. On a positive note, he singled out several pipeline, funding, and other initiatives. They included the Minority Medical Education Program (MMEP), started in 1989 by the Robert Wood Johnson Foundation

TABLE 5-1 Number and Percentage of U.S. Medical School *Applicants* in 1980 and 2016 by Race or Ethnicity

Race or Ethnicity	1980 Number	1980 Percent	2015 Number	2015 Percent
American Indian or Alaska Native	156	0.4	127	0.2
Asian	1,643	4.6	10,906	20.6
Black or African American	**2,507**	**7.0**	**4,344**	**8.2**
Hispanic or Latino	1,764	5.0	3,300	6.2
White	29,256	81.1	25,544	48.2
Total	36,083[a]		53,042[b]	

[a] Total includes 757 (2.1% of applicants) unknown and non-U.S. citizens and nonpermanent residents not included in the analysis

[b] Total includes 8,821 (16.6% of applicants) Native Hawaiian or other Pacific Islander, multiple-race, other, unknown, and non-U.S. citizens and nonpermanent residents not included in the analysis.

SOURCE: David Acosta, Workshop Presentation, April 13, 2020, from AAMC Data Warehouse: Applicant Matriculant File, archived January 2004.

(RWJF); Project 3000 by 2000, launched in 1991 by AAMC; and the Health Professions Partnership Initiative, launched by RWJF and the W.K. Kellogg Foundation in 1996 and 1998 to foster more involvement by medical schools in their local school communities to expand opportunities. In addition, the AAMC's Liaison Committee on Medical Education (LCME) issued two standards during this time: the LCME MS-8 in 2002 and the LCME IS-16 in 2009. This latter standard, Dr. Acosta said, strengthened the wording from a "you should" to a "you must" related to student and faculty diversity. According to Dr. Acosta, slight increases in applications by African Americans are shown from 1990 to 1995, then a downward trend from 1995 to 2005, then another slight increase after passage of the LCME IS-16. He also commented that Asian Americans seemed to benefit the most from these programs in terms of increases in applications.

Barriers during this time period included anti-affirmative action laws, beginning with Proposition 209 in California in 1996. Other states that have passed anti-affirmative action laws include Texas (1996), Washington State (1998), Florida (1999, by Executive Order), Michigan (2006), Nebraska (2008), Arizona (2010), Oklahoma (2012), New Hampshire (also 2012), and Idaho (2020). In 2019, a long-standing case involving admission policies at the Texas Tech University Health Sciences Center was resolved.[3] Dr. Acosta emphasized that affirmative action and actions against it are "very much alive," but a number of resources to help institutions document their admissions processes, including publications from AAMC (AAMC, 2014), the College Board and Education Counsel (Coleman and Keith, 2018), and Urban Universities for Health (2014). He also noted that the amicus briefs from the 2015 U.S. Supreme Court case Fisher v. University of Texas at Austin provides useful summaries of issues involved with affirmative action in admissions.[4]

Medical School Matriculation

In 1980, 999 Black and African American students matriculated in U.S. medical schools, and that number increased to just more than 1,400 in 2016 and 1,540 in 2018. The number of total seats rose from 16,587 in 1980 to 21,030 in 2015 (Table 5-2), and to 21,622 in 2018. Dr. Acosta characterized the figures:

> The percentage has gone from 6.0 percent in 1980 to 7.1 percent in 2016. It is amazing over that span of years, how low a number that is, particularly keeping in mind the total number of seats.

Dr. Acosta again pointed out that this number includes some people with a combination of races, as well as people born outside of the United States. Disaggregating, he pointed out, shows that 284 Black, U.S.-born males were enrolled in medical schools out of more than 21,000 seats.[5]

Looking at graduation rates among underrepresented minorities between 2001 and 2002, Blacks and African Americans had the lowest 4-year rate at 69.6 percent, although the rate increased after 5 and 6 years to a comparable level with other groups. He stated,

> Questions remain. We need further research to find out why the rate is so low after 4 years. Instead of going by anecdotal experience, we need the research and the data to home in what we can possibly highlight and change moving forward.

---

[3] For background, see Coleman and Keith (2019).

[4] A compilation of the documentation related to this case can be found at https://legal.utexas.edu/scotus-2015 [Accessed May 14, 2020].

[5] For more background on Black males in medicine, see NASEM (2018).

TABLE 5-2 Number and Percentage of U.S. Medical School *Matriculants* in 1980 and 2016 by Race or Ethnicity

Race or Ethnicity	1980 Number	1980 Percent	2015 Number	2015 Percent
American Indian or Alaska Native	63	0.4	54	0.3
Asian	679	4.0	4,475	21.3
Black or African American	**999**	**6.0**	**1,497**	**7.1**
Hispanic or Latino	807	4.9	1,335	6.3
White	13,884	83.7	10,828	51.5
Total	16,587[a]		21,030[b]	

[a] Total includes 155 (9% of matriculants) unknown and non-U.S. citizens and nonpermanent residents not included in the analysis.

[b] Total includes 2,841 (13.5% of matriculants) Native Hawaiian or other Pacific Islander, multiple-race, other, unknown, and non-U.S. citizens and nonpermanent residents not included in the analysis.

SOURCE: David Acosta, Workshop Presentation, April 13, 2020, from AAMC Data Warehouse: Applicant Matriculant File as of August 22, 2017.

GRADUATE SCHOOL ENROLLMENT IN SCIENCE AND ENGINEERING

Turning to science and engineering disciplines, Dr. Acosta said graduate school enrollment mirrors the situation in medical schools. According to data compiled by the National Center for Science & Engineering Statistics of the National Society Foundation, the number of Blacks or African Americans earning science or engineering bachelor's degrees as a percentage of the overall degrees granted has been static since 1996, with the lowest numbers in engineering and the highest in psychology, social sciences, and computer sciences. The anti-affirmative action measures discussed above also had an effect in science and engineering fields. Overall, in 2016, about 4.9 percent of all science and engineering graduate students were Black or African American (Table 5-3). Disaggregated by sex, 6.7 percent of the total pool were African American/Black females and 3.6 percent are African American/Black males.

MEDICAL SCHOOL FACULTY REPRESENTATION

Medical school faculty are predominantly male, as Dr. Acosta showed with data compiled by AAMC. In 2010, there were 52,300 women faculty

TABLE 5-3 U.S. Science and Engineering Graduate Students by Field, Race, and Ethnicity, 2016

Total	African American/ Black	Hispanic	American Indian/ Alaska Native	Asian	White
620,489	**30,600** **(4.9%)**	39,578	1,860	35,674	237,563
Engineering					
168,443	**3,710** **(2.2%)**	6,966	245	9,920	45,622
Sciences					
452,046	**26,890** **(6.0%)**	32,616	1,615	25,772	191,194

SOURCE: David Acosta, Workshop Presentation, April 13, 2020, from the National Center for Science & Engineering Statistics, NSF 19-304.

in U.S. medical schools and 92,047 males. In 2019, there were more female faculty but a gender gap of about 27,600 positions in academic medicine remained.

By race and ethnicity, among full-time faculty, there were 113,794 whites; 35,726 Asian; 6,503 African American/Black; 5,929 Hispanic; and 458 American Indian/Alaska Native/Pacific Islander. There were more African American women than African American men: 3,806 full-time female faculty and 2,697 full-time male faculty. Men and women are about at par in the basic sciences, but the larger proportion of women come in the clinical sciences, which have the largest number of medical faculty positions overall. Most of the women are teaching, in order of frequency, pediatrics, internal medicine, and obstetrics/gynecology; men are most commonly in internal medicine, pediatrics, and surgery (Figure 5-1).

Looking at all underrepresented minority science and engineering Ph.Ds. employed in academia, Dr. Acosta characterized these numbers as "lackluster, with the needle barely moved along the way." Percentage-wise, African American and Black women filled 4.7 percent of full-time positions in 2003, and had slightly declined to 4.6 percent in 2013. African American and Black men filled 2.9 percent of the full-time faculty positions in 2003, a percentage that slightly increased to 3.3 percent in 2013.

FIGURE 5-1 U.S. medical school full-time faculty by race and ethnicity, 2019.
SOURCE: David Acosta, Workshop Presentation, April 13, 2020, based on AAMC data.

MANIFESTATIONS OF STRUCTURAL AND INSTITUTIONAL RACISM

With these data in mind, Dr. Acosta shared a cartoon in which a white man and African American woman are "racing" toward a finish line. The man's route has a few hurdles along the way; the woman's is filled with barbed wire, walls, and the like. Quoting a former senior academic editor at the American Association of Colleges and Universities, "Our institutional policies and practices, infrastructure, governance, unspoken rules, and expectations have perpetuated a double standard for specific groups." He added that intersectionality also plays a role and sometimes learners, faculty, and staff are confronting triple or even quadruple standards.

Several studies have identified challenges faced by medical students and residents from historically excluded and underrepresented groups in STEMM (HEUGS), Dr. Acosta said, including difficulties in acculturation to the culture of medicine, mistreatment, microaggressions, isolation and marginalization, racial biases, prejudice and discrimination, and the imposter syndrome. Dr. Acosta stated:

> We haven't dealt successfully from a systemic standpoint with these challenges—looking at the infrastructure, the policies, the practices that have perpetuated these challenges, and I think those are going to have to be addressed as we begin to find solutions that are systems-based. We need to move beyond blaming the typical things we tend to blame; I think it's a higher order now that we have to begin looking at.

Each year medical students fill out an AAMC survey about their medical school experience. About 40 percent of those who responded to the most recent survey reported personal experience with harassment or treatment that included racial discrimination.[6] Other studies have similar findings. A 2011 systematic review of the literature showed 59.4 percent had experienced at least one form of harassment or discrimination during their training, and about 26 percent indicated they had at least one experience with racial discrimination (Fnais et al., 2014). Medical school faculty from HEUGS reported a number of challenges, including difficulties in acculturation, racial battle fatigue, sexual harassment, and inequity in salaries and promotion.

---

[6] See findings of the AAMC 2019 Graduation Questionnaire at https://www.aamc.org/system/files/reports/1/2018gqallschoolssummaryreport.pdf [Accessed May 14, 2020].

The impacts, he said, may include poor emotional and mental health outcomes, post-traumatic stress disorder, and burnout. Burnout is an issue for many medical faculty members, as it is for physicians (Dandar, Grigsby, and Bunton, 2019). The percentages of women (of all races and ethnicities) who reported they were burning out or were already burnt out are higher than for men.

Dr. Acosta concluded that the numbers and percentages of African American and Black students applying to medical school and science and engineering graduate schools have remained relatively flat over the past 30 to 40 years, despite multiple efforts to address the problem. Anti-affirmative action laws have impacted these efforts to accelerate the enrollment. The number of African American and Black matriculants and faculty have also remained relatively flat, and it is important to disaggregate the data and interpret them correctly. He also urged more research to learn about graduation rates, students who take the Medical College Admission Test (MCAT) but do not apply to medical school, where African American physicians practice and for whom they provide patient care, and the disciplines in which enrollment is particularly low, such as engineering. Finally, it is important to look carefully at institutional environments. Racial tension, biases, and discrimination are on the rise and are impacting learners and faculty, he concluded.

DISCUSSION

In response to a question from a participant about any differences in graduation rates between historically Black compared to predominantly white medical schools, Dr. Acosta said that AAMC and Accreditation Council for Graduate Medical Education have conducted a survey to focus on attrition in residency programs. They are looking upstream to see the institutions from which the residents come, which might help answer this question. Findings are expected in the next 3 to 6 months.

The issue of standardized tests and their role in medical school admissions and the residency programs was raised. Clarifying that he was speaking as an individual and not for AAMC, he suggested that when the U.S. Medical Licensing Examination (USMLE) becomes pass-fail, it may level the playing field; for a long time, the "Step 1" scores have functioned more as a screening-out than a screening-in process. He suggested this change may open up a discussion on how to look at a student's distance traveled and other attributes, as well as relying less on MCAT scores for medical school admission. Osteopathic medicine moved away from the MCAT for admissions in 2020 because of COVID-19, a decision that may provide an example for content experts to study and discuss, he suggested. He also called attention to the students who do not "match" a residency program because of test scores, noting he himself did not pass Step 1 of the USMLE until his third attempt. "I worry about the students who get lost in the system but should remain because they have such talent," he said. "The good news is that it starts with conscious awareness of a problem to get to a solution."

Dr. Jones further suggested finding a way to follow up with individuals to learn and communicate about the lost talent that results. "I think of all the people who took the MCAT but never applied to med school," she said. "We act as if that is all right, but we don't understand the magnitude of the loss in terms of lost talent." Dr. Acosta agreed this could be an important area of research. He added it is important to communicate more about the accomplishments of students who overcame significant adversity and succeeded.

Related to admissions, Dr. Acosta pointed to the need to look at members of admissions committees. "Their performance needs to be evaluated," he asserted. "How did they screen in or screen out students? What were their scores? How many did they advocate for?" There may be a need for term limits or not retaining members who cannot be trained in this broader view. He said that about 25 percent of admissions committees include community members. Some states have more control over universities than others in setting the bar and limiting entry for underrepresented minorities. There is a need to identify and be vocal about those states and state legislatures, Dr. Acosta said.

Referring to Dr. Acosta's data related to application and enrollment numbers, a participant asked why bridge programs seem not to have a large impact on the numbers. Dr. Acosta replied,

> The problem is not the pipeline, the problem is who is in control to get past the gate into med schools.... We have a body of people [medical schools] who control what our workforce looks like. We have the data because people ask for the data, we've created pipeline programs because people have asked for them to meet accreditation, we have donations from philanthropic organizations who believe in and see the value in it. The reality is we have done a lot of this work and still this [disparity] prevails.... It's about essentially there are laws, there are policies, there are practices, there are values and customs and beliefs specifically not to upset the status quo.

He also noted the need for data about where students in pipeline programs are accepted; anecdotally, he commented that many schools have excellent pipeline programs but the students in them have to go elsewhere for admission to medical school. Quantitative data are also needed to validate whether increasing the number of African American faculty would be reflected in a more diverse talent pool. Expanding the number of underrepresented minorities in leadership positions might also help, he added. To that end, AAMC conducts several programs, including minority faculty leadership summits for early career and mid-career faculty. For the past 5 years, a health executive diversity certification program has been offered for assistant and associate deans, as well as current and potential chief diversity officers.

Dr. Acosta acknowledged that different states have different demographic compositions in terms of who makes up that state's minorities or disadvantaged groups. He urged, "In defining diversity for your institution, define it as you will, but there's an important question you have to answer, too. How is your institution addressing the national crisis based on the data we have for all population groups?"

REFERENCES

Association of American Medical Colleges (AAMC). 2014. Roadmap to Diversity and Educational Excellence: Key Legal and Educational Policy Foundations for Medical School, Second Edition. Prepared for AAMC by A. L. Coleman, K. E. Lipper, T. E. Taylor, and S. R. Palmer. https://store.aamc.org/downloadable/download/sample/sample_id/192.

Coleman, A., and J. L. Keith. 2018. Understanding Holistic Review in Higher Education Admissions: Guiding Principles and Model Illustrations. Prepared for College Board and Education Counsel. https://professionals.collegeboard.org/pdf/understanding-holistic-review-he-admissions.pdf.

Coleman, A., and J. L. Keith. 2019. Race in admissions in the wake of the Texas Tech resolution. https://www.aamc.org/news-insights/insights/race-admissions-wake-texas-tech-resolution.

Dandar, V., R. K. Grigsby, and S. Bunton. 2019. Burnout among faculty in academic medicine. *Analysis in Brief,* 19(1), 1–2.

Fnais, N., C. Soobiah, M. H. Chen, E. Lillie, et al. 2014. Harassment and discrimination in medical training: A Systematic review and meta-analysis. *Academic Medicine,* 89(5), 817–827.

National Academies of Sciences, Engineering, and Medicine (NASEM). 2018. *An American Crisis: The Growing Absence of Black Men in Medicine and Science: Proceedings of a Joint Workshop.* Washington, DC: The National Academies Press. https://www.nap.edu/catalog/25130/an-american-crisis-the-growing-absence-of-black-men-in.

Steinecke, A., and C. Terrell. 2010. Progress for whose future? The impact of the Flexner Report on medical education for racial and ethnic minority physicians in the United States. *Academic Medicine,* 85, 236–245.

Urban Universities for Health. 2014. Holistic Admissions in the Health Professions: Findings from a National Survey. https://urbanuniversitiesforhealth.org/media/documents/Holistic_Admissions_in_the_Health_Professions_final.pdf.

6

Racism in Relation to Black Men and Women in Science, Engineering, and Medicine

---

Highlights from the Presentation

- The COVID-19 crisis serves as way to highlight the impact of systemic bias, with a large gap in mortality rates between whites and African Americans.
- Major factors that underlie systemic bias, as highlighted with COVID-19 mortality rates, include socioeconomic factors; health disparities; bias, assumptions, and structural racism; everyday logistics; and psychological impacts.
- Changing systems requires thinking from a systems perspective; without a systems view, individual actions may even have a long-term negative effect.
- In times of crisis, fear and "otherization" are exacerbated.
- Interventions against structural racism include those related to education, priming, structures and systems, and accountability.

Howard Ross, founder of Cook Ross and author of several publications on diversity and bias, was the final presenter before the workshop's concluding discussion. He used the COVID-19 crisis to illuminate the impact of systemic bias, pointing to the large difference in mortality rates between whites and African Americans from the disease. He identified five major factors that underlie this gap: socioeconomic factors; health disparities; bias, assumptions, and structural racism; everyday logistics; and psychological impacts. "What I want to weave here is a tapestry that a lot of seemingly disconnected factors play a dramatic role in the numbers related to the Black-white disparity in mortality rates related to COVID-19," he said.

FACTORS UNDERLYING THE COVID-19 MORTALITY GAP

Socioeconomic Factors

As noted earlier in the workshop, African Americans have 7 percent the acquired wealth of white households, making for less of a nest egg to fall back on. African Americans also have a lower percentage of salaried versus hourly jobs, and Black unemployment (even before COVID-19) was roughly twice as high as white unemployment. Roughly one in six Black and Hispanic households spend more than 50 percent of their income on housing. The impact is a greater incentive to work, less economic reserves, and less access to food, as well as less access to medical support and funds to buy personal protective equipment like masks and gloves. African Americans also have a higher percentage of "essential" jobs that bring higher exposure to the coronavirus to themselves and their families.

Health Disparities

Statistics compiled by the Centers for Disease Control and Prevention present data that show gaps in health between African American and white populations in the United States.[1] Mr. Ross reported that these data show that African Americas are 50 percent more likely to suffer from high blood pressure, 65 percent more likely to suffer from diabetes, 75 percent more likely to have a stroke, 50 percent more likely to have some form of cardiovascular disease, and 60 percent more likely to suffer from childhood asthma. African Americans also have more than twice the infant mortality rate and have significantly more cases of HIV infection. According to data that Mr. Ross cited from a Georgetown University study, African Americans in Washington, DC, are 3.5 times more likely to live below the poverty level and also are more likely to suffer or die from a number of health conditions (Georgetown University School of Nursing & Health Studies, 2016). These

---

[1] The U.S. Centers for Disease Control and Prevention maintains the National Vital Statistics System at https://www.cdc.gov/nchs/nvss/index.htm [Accessed May 14, 2020].

conditions, in addition to health disparities brought on by such environmental problems as climate change and exposure to pollutants and toxins, are co-morbidities that are directly impacted by and make people more vulnerable to COVID-19, Mr. Ross said.

Biases, Assumptions, and Structural Racism

Historical narratives about race have created structural arrangements (e.g., where people live and work) and differential treatment that create a dominant/non-dominant relationship between whites and people of color, especially African Americans, Mr. Ross stated. He noted these structures and systems were built many years ago, and they have evolved, and often devolved, over time. He stated:

> One of the challenges we have is that so much of it [inequality] is what we call our "normal" existence that many people do not question or challenge it. We see people in circumstances and rather than judge the circumstance they are in, those in the dominant community tend to judge the people in the circumstance.... The victim becomes seen as the cause of their own victimization.

Policies and practices reinforce the narrative, and unconscious bias impacts African Americans in many ways, especially in health care, he said, referring to several studies to illustrate his points. An empirical analysis of racial differences in police use of force (Fryer, 2016, 2018) showed that white police officers were more likely to use violence against Black compared with white members of the public. These actions, he said, are tied to "who looks dangerous" to a police officer and can be visceral, fast-brained reactions. Connected with the COVID-19 crisis, he pointed to a disproportionate number of African American men in prison, where social distancing and other preventative health measures are difficult to implement.

Other research points to implicit bias in the provision of health care. For example, a study revealed how implicit bias among physicians can predict their thrombolysis decisions (related to procedures to treat blood clots) for their Black and white patients (Green et al., 2007). A literature review published in the *American Journal of Public Health* (Hall et al., 2015) showed implicit racial/ethnic bias among health care professionals in 14 out of the 15 studies under review. Thus, he said:

We are not talking about something that happens every once in a while, where you happen to have an errant person…. We are talking about a system producing this result time after time.

Related to higher education, Mr. Ross described a study in which researchers pretending to be students sent emails to 6,500 professors in 89 disciplines at the top 259 schools in the country (Milkman, Akinola, and Chugh, 2015). The names of the students were chosen to express gender and ethnicity. The emails signed with what seemed like white male names were 25 percent more likely to receive a response from a professor, with the percentage rising among professors at the highest-rated schools and in higher salaried subjects. Ross said:

> Responsiveness from your professor, the ability to get communication, to get information, to get your questions answered—when the professor says, 'I have an open door policy,' is it really open to everybody? All of these are factors which impact students' ability to be successful and contribute to the relationships that get built that are so important. And it's also important for us to recognize that these mentoring relationships, these sponsorship relationships tend to be much more important for people in non-dominant groups than in dominant groups.

A study of the membership in the Alpha Omega Alpha Honor Society showed that students of color were significantly less likely to be chosen as members, even when controlling for other factors: Blacks were less than 20 percent as likely as white students to become accepted into these societies, Asians were 50 percent as likely, and Latinx students were 79 percent as likely (Boatright et al., 2017). Many medical school programs use these memberships as a formal or informal criterion for acceptance, Mr. Ross pointed out.

These issues regularly happen across society, he continued, but fear, such as that surrounding COVID-19, exacerbates the impact. The collective human mind undergoes what Daniel Goleman has called an "amygdala hijacking." When in extreme trauma, this "fear center" part of the brain takes over, which can result in reactive responses (fight, flight, or freeze); protective responses (desire for control, a turn toward authoritarianism, or increased "otherization"); and narrowed thinking and interpretations (preventing a more collective response).

Everyday Logistics

Many difficult everyday logistics disproportionally impact African American communities, Mr. Ross stated. These factors include accessibility to transportation, health care, and safe markets and commercial spaces; lower levels of health insurance; higher levels of pollution; lack of green spaces and recreational facilities; closer physical proximity; and lack of trust in law enforcement.

Psychological Impact

These factors taken together create a psychological impact that has health implications, as shown with COVID-19. Mr. Ross referred to psychological safety, with feelings of increased stress and vulnerability; questionable trust in government; limited access to information; less trust in medical research; learned helplessness; and "otherization" and perceptual safety. He quoted an African American man, writing in the *Guardian* newspaper, who expressed fear about wearing a face mask while going to the grocery store despite the public health guidance to do so (Thomas, 2020). As he wrote, "my voice of self-protection reminded me that I, a Black man, cannot walk into a store with a bandanna covering the greater part of my face if I also expect to walk out of that store…. For me, the fear of being mistaken for an armed robber or assailant is greater than the fear of COVID-19."

SYSTEMIC IMPACTS AND NEED FOR CHANGE

Mr. Ross agreed with previous presenters that the systemic impact of racism requires systemic change. An example of a systemic change relates to the gender composition of major symphony orchestras. Orchestras were 5 percent women in 1970 and 12 percent women in 1980. When changes were made—such as openly announcing auditions and conducting them behind a screen so the evaluators would hear the quality of the music but not see the person performing—the numbers rose to more than 40 percent.

Changing systems requires thinking from a systems perspective, he said. Moreover, he added, without this systemic view, individual actions may have a negative long-term effect. He urged interventions in four domains to create systems that are less biased: education, priming, systems and structures, and accountability.

In terms of education, Mr. Ross acknowledged disagreements about whether training to reduce or eliminate unconscious bias is effective but asserted that bias awareness training can work. One study showed that this type of training can help medical professionals "learn about unconscious processes that provide them with skills that reduce bias when they interact with minority group patients," he said, quoting from a study by Stone and Moskowitz (2011).

Priming refers to follow-up after training that can help people "at the moment it is needed," he explained. For example, a person on a selection committee can participate in bias awareness training but not use the information for several months. Showing a short video or providing a series of questions based on the training right before going into a selection committee meeting increases the impact of the training, he said. Related to this concept is "nudging" to embed certain behaviors. The example he provided is when people "opt out" of designating they would like to be an organ donor versus "opting in." Looking at European countries, the opting-out system results in near-universal organ donor designation, while the opting-in system results in far lower participation. Mr. Ross particularly called out the impact of a nudge when looking at two countries with similar cultures: Austria (99.98 percent organ donors with an opt-out policy) compared to neighboring Germany (12 percent with an opt-in policy). Related to COVID-19, priming and nudging examples include signs that show where to stand to keep the recommended distance from others. More broadly, there are things that can be built into systems to remind people on a daily basis about what they are supposed to be doing.

Looking at aligning structures and systems, Mr. Ross focused on the workplace, such as removing names from the initial screening of resumes, checking interview questions for bias, using diverse panels for hiring, and conducting fair performance reviews. "In every area of talent management, we can find ways that bias impacts us," he said. "None of these individually is going to change the system, but the collection of all of them has a huge impact on the system."

The fourth domain is accountability. In looking at the many parts of a system, he stressed the need for metrics to determine if differential impacts are occurring related an organization's history, values, and the environment.

DISCUSSION

Harriet Washington referred to the work of social psychologist Mark Schaller, who has proposed that protective prejudice and heightened degrees of xenophobia come into play during a crisis, which totalitarian regimes can capitalize on. Mr. Ross agreed with this theory, as shown with excessive otherization. Both otherization and production of empathy can now be detected in images of the brain.

Another participant asked how to counter the claims of some powerful whites that people of color are genetically inferior as a scapegoat to acknowledge implicit bias. Mr. Ross stated it is the primary responsibility of the white community, not just people of color, to raise awareness and correct this claim. Those in the dominant group, he said, often do not pay attention to their own privilege and entitlement. Social structure also reinforces this, with some people believing that "if there is someone below me, I feel better."

Dr. Jones referred to her keynote address (see Chapter 2), in which she pointed out that one of the impacts of racism is to sap the strength of the whole society. She asked Mr. Ross and the entire group how to better communicate that "racism is for real and is sapping the strength of the whole society through the wasting of human resources." Mr. Ross responded, "We will have no solution to these problems until we see them as collective, societal problems and not the problems of individual groups. This is one reason why our ability to establish the business case, the financial case, the economic case, the performance-based case for everyone's full participation is so important."

Mr. Ross said he uses a business-case approach in diversity, equity, and inclusion (DEI) work with organizations, in addition stating moral convictions. "I know some people are sick and tired of building the business case," he acknowledged, "but it's amazing how many people still don't get it, and also how it continues to evolve." As the nation's demographic composition changes, he added, it becomes easier to make that case with leaders, stressing the impact on "your business, your hospital, your patient outcomes, turnover." While he suggested speaking to "the goodness in people about the importance of diversity, we have to deal with reality, and go in where the door is open." He said a business case can be found in almost all situations, although sometime requires some digging.

Ms. Washington pointed out that in the medical sphere, medical interdependence and the intersection of risk is often a compelling argument to find ways to reduce disparities. Environmental toxins have been shown to affect mental capacity as measured by IQ scores, predominately in communities of color but also the nation as a whole, she added, while noting the debate about what IQ tests truly measure.

Turning to assimilation, Dr. Bright observed that the minority culture is usually expected to change to adapt to the dominant culture and asked how that can change. Mr. Ross observed that most people in the dominant culture are so used to this, they do not see that assimilation has even taken place. Environments are needed to explore other cultures without cultural appropriation, he said.

In terms of increasing the number of Black men and Black women in medicine, Dr. Jones suggested a strategy for medical school committees to "burst through their bubbles of experience" by creating opportunities to experience common humanity in different settings and join together in common cause. Related to residential segregation, she said, "the most profound damage is that the walls are so tall and yet invisible—people don't even know there is another reality." Examples suggested including pairing science classes from different schools to work on a common project, a multicultural Passover Seder, and a series of "equitable dinners" in Atlanta, Georgia. Regular organizational activities, such as setting up mixed culture teams on projects or book clubs, were also cited as examples.

Health disparities have an economic cost as well. David Acosta elaborated on what he termed the cost effectiveness of health disparity work and its long-term value. Diversity officers and others can show that a diverse workforce improves health outcomes and saves money through working with the community, reducing hospital re-admittance, and increasing health literacy, decreasing the number missed appointments, or reducing errors from not understanding health-related instructions. Dr. Jones urged mobilizing to address the seven barriers to the achievement of health equity (see Chapter 2) as values targets, along with tackling structural targets such as residential segregation, to develop a more complete approach to an anti-racism agenda.

The session ended with a discussion of impacts on African Americans with higher income levels. In addition to the historical reasons discussed, many may experience a wealth gap because of societal expectations, suggested one participant. Cato Laurencin referred to two studies: one that showed that the son of a Black millionaire and the son of a white single mother earning less than $36,000 a year have statistically the same chance of becoming incarcerated. Another study measured the allostatic load among

African Americans, regardless of income level, incurred through such daily activities as driving with an increased likelihood of being stopped by a police officer.

A participant asked about countering anti-affirmative action efforts in medical schools. Dr. Acosta said much can be learned from institutions, such as University of Washington, University of California, and University of Michigan, that have gone through the process (see also Chapter 5). It is important to develop the documentation to interpret the law correctly and not over-interpret the laws.

REFERENCES

Boatright, D., D. Ross, P. O'Connor, et al. 2017. Racial disparities in medical student membership in the Alpha Omega Alpha Honor Society. *JAMA Internal Medicine*, 177(5), 659–665. DOI: 10.1001/jamainternmed.2016.9623.

Fryer, R. G., Jr. 2016, Revised 2018. An Empirical Analysis of Racial Differences in Police Use of Force. Working Paper 22399, National Bureau of Economic Research. http://www.nber.org/papers/w22399.

Georgetown University School of Nursing & Health Studies. 2016. *The Health of the African American Community in the District of Columbia; Disparities and Recommendations.* Prepared for the D.C. Commission on African American Affairs. https://www.abfe.org/wp-content/uploads/2016/11/The-Health-of-the-African-American-Community-in-the-District-of-Columbia.pdf.

Green, A. R. et al. 2007. Implicit bias among physicians and its prediction of thrombolysis decisions for Black and white patients. *Journal of Internal Medicine,* 22(9), 1231–1238. doi: 10.1007/s11606-007-0258-5.

Hall, W. J. et al. 2015. Implicit racial/ethnic bias among health care professionals and its influence on health care outcomes: a systematic review. *American Journal of Public Health*, 105(12), e60–e76.

Milkman, K. L., M. Akinola, and D. Chugh. 2015. What happens before? A field experiment exploring how pay and representation differentially shape bias on the pathway into organizations. *Journal of Applied Psychology, 100*(6), 1678–1712. https://doi.org/10.1037/apl0000022.

Stone, J., and G. Moskowitz. 2011. Non-conscious bias in medical decision making: What can be done to reduce it? *Medical Education* 45(8), 768–776. doi: 10.1111/j.1365-2923.2011.04026.x.

Thomas, A. 2020, April 7. I'm a Black man in America. Entering a shop with a face mask may get me killed. *The Guardian.* https://www.theguardian.com/commentisfree/2020/apr/07/black-men-coronavirus-masks-safety.

7

Concluding Remarks

Dr. Cato T. Laurencin, chair of the Roundtable, expressed appreciation to the presenters and participants. Dr. Camara Jones, workshop co-chair, then presented closing remarks, drawing from some of the points she made during her keynote address (see Chapter 2).

Dr. Jones challenged the participants with what she termed homework. She asked them to view how racism may be operating in different places and contexts with which they have contact and identify levers of interventions to counter it. She noted she had developed some "top-of-mind" suggestions related to each mechanism and asked participants to identify others.

As she presented in the keynote, Dr. Jones said mechanisms that contribute to Black underrepresentation in medicine may include:

- Structures, such as racial residential segregation, quality preschool programs, K-12 partnerships, or medical faculty diversity
- Policies, such as unequitable public school funding or truncated affirmative action policies
- Practices, including use of MCAT scores as a first hurdle, inclusion of photographs on applications, and limited outreach to HBCUs
- Norms, including that standardized tests are important predictors of success as a clinician while a student's "distance traveled" is marginally relevant
- Values, including beliefs that Blacks are less intelligent or hard-working or other negative traits, and white supremacist ideology.

"This is not just a stimulating day with great talks," Dr. Jones said. "We want this Roundtable meeting to create a menu of levers for dismantling racism and putting in its place a system where we can all develop to our full potential, including the genius of Black men and Black women in science, engineering, and medicine."

"As the COVID-19 pandemic winds down, we cannot return to the status quo. We need a new normal," Dr. Jones said. She called for three items on an anti-racist agenda: "reparations, de-carceration, and a huge investment in communities with a special focus on investment in children."

Workshop co-chair Cedric Bright closed by noting the workshop discussions covered racism and bias as a multifaceted issue, and the work of the Roundtable would continue. "This is not the end, this is just the beginning," he said.

Appendix A

Workshop Agenda

The Impacts of Racism and Bias on Black People Pursuing Careers in Science, Engineering, and Medicine: A Workshop

April 13–14, 2020 at 9:30 AM EST

Virtually via ZOOM

9:30 AM **Opening Remarks**
Victor Dzau, M.D., President, National Academy of Medicine
Vaughan Turekian, Ph.D., Executive Director, Policy and Global Affairs
Cato T. Laurencin, M.D., Ph.D., Chair of the Roundtable
Camara Phyllis Jones, M.D., M.P.H., Ph.D., Co-Chair, Racism and Bias Action Group
Cedric M. Bright, M.D., FACP, Co-Chair, Racism and Bias Action Group

10:00 AM **Keynote**
Camara Phyllis Jones, M.D., M.P.H., Ph.D. – Newest Aspects of Racism

10:45 AM - 11:30 AM	**Harriet A. Washington** *Medical and Public Health History* • What is the history of the involvement of "Black" people in medicine? • What is the history of the involvement of "Black" people in public health? • What is the history of proposing health care as a right in this country?
11:30 AM - 12:00 PM	**Questions and Answers**
12:05 PM - 12:50 PM	**Richard Rothstein** *Federal Legislation Segregating Housing and Education* • What is the history of the "Black"-"white" wealth disparity over time in this country? • What mechanisms are in place that perpetuate initial historical injustices? • What is the impact of the GI bill on housing segregation and the resultant educational, employment, environmental, health care, and other segregation? • What is the impact of funding public schools based on local property taxes? • Why is access to excellent public education not a right guaranteed by our Constitution?
12:50 PM - 1:05 PM	**Questions and Answers**
1:05 PM - 1:50 PM	**Lunch**
1:50 PM - 2:35 PM	**David A. Acosta, M.D.** *Attacks on Diversity/Equity/Inclusion in Education.* • What is the historical trend of the enrollment of "Black" people in medical schools? Engineering schools? The sciences?

	• What has been the impact of the Flexner Report on these numbers? The Bakke decision? The Fischer decision? Other court challenges and Supreme Court decisions regarding the use of "race" in admissions?
• What is the historical trend of "Black" faculty representation in medicine? Engineering? The sciences?
2:35 PM - 3:20 PM	**Questions and Answers**
3:25 PM - 4:10 PM	**Howard Ross**
Racism in Relation to Black Men and Black Women in Science, Engineering, and Medicine
• What is racism?
• How does it relate to bias?
• How does it structure opportunity?
• How does it assign value?
• How does it sap the strength of the whole society?
• How can implicit and explicit bias training mitigate the impacts of these biases on "Black" people?
• Does implicit and explicit bias training combat the effects of racism? If not, what else is needed?
4:10 PM - 4:25 PM	**Questions and Answers**
4:25 PM	**Final Questions – Moderated by Dr. Jones**
4;50 PM	**Concluding Remarks from Co-Chairs**
5:00 PM	**Adjourn**

Appendix B

Biographical Sketches of Roundtable Members and Workshop Presenters

BIOGRAPHIES OF ROUNDTABLE MEMBERS

Cato T. Laurencin (*Chair*) is the eighth designated University Professor in the 135-year history of the University of Connecticut. He is the Albert and Wilda Van Dusen Distinguished Endowed Professor of Orthopaeic Surgery. He is also the chief executive officer of The Connecticut Convergence Institute for Translation in Regenerative Engineering and the director of the Raymond and Beverly Sackler Center for Biomedical, Biological, Physical, and Engineering Sciences at the University of Connecticut. Dr. Laurencin earned a B.S.E. in chemical engineering from Princeton University, and his M.D., magna cum laude, from the Harvard Medical School, and received the Robinson Award for Surgery. He earned his Ph.D. in biochemical engineering/biotechnology from the Massachusetts Institute of Technology, where he was named a Hugh Hampton Young Fellow. A practicing sports medicine and shoulder surgeon, Dr. Laurencin has been named to America's Top Doctors for more than 15 years. He is a fellow of the American Academy of Orthopaedic Surgeons, a fellow of the American Orthopaedic Association, a fellow of the American College of Surgeons, and a member of the American Surgical Association. He received the Nicolas Andry Award, the highest honor of the Association of Bone and Joint Surgeons. Dr. Laurencin served as dean of the Medical School and vice president for health affairs at the University of Connecticut. Dr. Laurencin is a pioneer of the new field, regenerative engineering. He is an expert in biomaterials science, stem cell technology and nanotechnology, and was named one of the 100 Engineers of the Modern Era by the American Institute of Chemical Engineers, and received the Founder's Award from the Society for Biomaterials. Dr. Laurencin received the National Institutes of Health (NIH) Director's Pioneer Award, the NIH's highest and most prestigious research award, for his new field of regenerative engineering, and the National Science Foundation's Emerging Frontiers in Research and Innovation Grant Award. Dr. Laurencin is the editor-in-chief of *Regenerative Engineering and Translational Medicine*, published by Springer Nature, and is the founder of the Regenerative Engineering Society. He is a fellow of the American Chemical Society, a fellow of the American Institute of Chemical Engineers, a fellow of the Biomedical Engineering Society, a fellow of the Materials Research Society, and a fellow of the American Association for the Advancement of Science (AAAS). The AAAS awarded Dr. Laurencin the Philip Hauge Abelson Prize given "for signal contributions to the advancement of science in the United States." Dr. Laurencin is active in mentoring, especially underrepresented minority students. He received the AAAS Mentor Award, the Beckman Award for Mentoring, and the Presidential Award for Excellence in Science, Mathematics, and Engineering Mentoring in ceremonies at the White House. The Society for Biomaterials established the Cato T. Laurencin, M.D., Ph.D. Travel Fellowship in his honor, awarded to underrepresented minority students pursuing research. Dr. Laurencin is also active in addressing health disparities. He completed the program in African-American Studies at Princeton University and is a core faculty member of the Africana Studies Institute at the University of Connecticut, and serves as editor-in-chief of the *Journal of Racial and Ethnic Health Disparities*, published by Springer Nature. He co-founded the W. Montague Cobb/NMA Health Institute, dedicated to addressing Health Disparities, and served as its founding chair. The W. Montague Cobb/NMA Health Institute and the National Medical Association established the Cato T. Laurencin Lifetime Research Achievement Award, given during the opening ceremonies of the National Medical Association Meeting. Dr. Laurencin is an elected member of the National Academy of Medicine, an elected member of the National Academy of Engineering, and an elected member of the American Academy of Arts and Sciences. Active internationally, he is an elected fellow of the Indian National Academy of Sciences, the Indian National Academy of Engineering, the African Academy of Sciences, The World Academy of Sciences, and is an Academician of the Chinese Academy of Engineering.

Olujimi Ajijola is an assistant professor in the departments of medicine-cardiology, and molecular, cellular, and integrative physiology at the University of California, Los Angeles (UCLA). Dr. Ajijola received his B.A. from the University of Virginia, his medical degree from Duke University, and his Ph.D. in molecular, cellular, and integrative physiology from UCLA. His clinical training in internal medicine and cardiovascular diseases/cardiac electrophysiology took place at the Massachusetts General Hospital/Harvard Medical School and at UCLA, respectively. Dr. Ajijola's clinical and research interests revolve around innovative methods to control life-threatening cardiac arrhythmias by modulating the autonomic nervous system. He is also an alumnus of the Howard Hughes Medical Institute's Medical Fellows Program, a recipient of the National Institutes of Health Director's New Innovator Award (DP2), and a Young Physician Scientist Award from the American Society for Clinical Investigation.

Mark Alexander is a retired research scientist at the Division of Research at Kaiser Permanente Northern California. He is the former Assistant Director of the Medical Effectiveness Research Center for Diverse Populations, University of California, San Francisco. Dr. Alexander is an epidemiologist who is committed to improving health outcomes of marginalized communities. The effects of racism and social class on health are of particular interest to Dr. Alexander. He is the National Secretary of 100 Black Men of America, Inc., and is a leader of the organization's Health and Wellness Committee. He is also an advisor to numerous community organizations in the San Francisco Bay Area. Dr. Alexander has published in the *Archives of Internal Medicine*, the *Journal of the American Medical Association*, the *American Heart Journal*, and other peer-reviewed journals. Dr. Alexander's research interests include cardiovascular disease, geriatric epidemiology, and child health. He is currently the executive director of Youth Movement, a community-based initiative dedicated to improving the health, fitness and wellbeing of Black children. Dr. Alexander received his bachelor's degree in biology from the University of California, Santa Cruz, his Master of Public Health from the University of California, Berkeley, and his Ph.D. in epidemiology from the University of California, Berkeley.

Gilda A. Barabino is the Daniel and Frances Berg Professor and Dean of The Grove School of Engineering at The City College of New York (CCNY). She holds appointments the in the departments of biomedical engineering and chemical engineering and the CUNY School of Medicine.

Prior to joining CCNY, she served as associate chair for graduate studies and professor in the Wallace H. Coulter Department of Biomedical Engineering at Georgia Tech and Emory University. At Georgia Tech she also served as the inaugural vice provost for academic diversity. Prior to her appointments at Georgia Tech and Emory, she rose to the rank of full professor of chemical engineering and served as vice provost for undergraduate education at Northeastern University. She is a noted investigator in the areas of sickle cell disease, cellular and tissue engineering, and race/ethnicity and gender in science and engineering. Dr. Barabino received her B.S. in chemistry from Xavier University of Louisiana and her Ph.D. in chemical engineering from Rice University. She is an elected member of the National Academy of Engineering and an elected fellow of the American Association for the Advancement of Science (AAAS), the American Institute of Chemical Engineers (AIChE), the American Institute for Medical and Biological Engineering (AIMBE), and the Biomedical Engineering Society (BMES). She is past president of BMES and past president of AIMBE. Her many honors include an honorary degree from Xavier University of Louisiana, the Presidential Award for Excellence in Science, Mathematics and Engineering Mentoring, and the Pierre Galetti Award, AIMBE's highest honor. Dr. Barabino is a trustee of Xavier University of Louisiana and a member of the National Science Foundation's Advisory Committee for Engineering, the congressionally mandated Committee on Equal Opportunities in Science and Engineering and the National Academies Committee on Women in Science, Engineering, and Medicine. She has served on the National Institutes of Health's National Advisory Dental and Craniofacial Research Council, and the National Academies Committee on the Impact of Sexual Harassment in Academia. Dr. Barabino consults nationally and internationally on STEM education and research, diversity in higher education, policy, workforce development, and faculty development. She is the founder and executive director of the National Institute for Faculty Equity.

Cedric M. Bright, M.D., FACP, is the associate dean for admissions, interim associate dean for diversity affairs and a full professor of internal medicine at the Brody School of Medicine at East Carolina University. He is a member of Alpha Omega Alpha, the Order of the Golden Fleece, and served as the 112th president of the National Medical Association from 2011 to 2012. He has previously held academic appointments at the University of North Carolina, Duke University, the Durham VA, and Brown University. Dr. Bright also was the chair the board of directors at the Lincoln Community Health Center as well as the Boys and Girls Club of Durham and Orange counties; has spoken about health disparities at the White House and before the Congressional Black Caucus; was a medical ambassador to Ghana; and has served as a mentor for the Student National Medical Association. He is a dedicated leader in delivering patient equity through broader access, and is a staunch proponent of health care reform.

L. D. Britt, a proud native of Suffolk, Virginia, and a member of the National Academy of Medicine, has strong southern roots and is the product of the public school system. He attended the University of Virginia and was named to the Dean's List each of his eight semesters. He received his B.A. with distinction. Dr. Britt, a graduate of Harvard Medical School and Harvard School of Public Health, is the Brickhouse Professor and chairman of the Department of Surgery at Eastern Virginia Medical School. He is the author of 300 scientific publications (with more than 231 being peer-reviewed articles). In addition, he is the author of three textbooks, including a recent edition of the highly touted *Acute Care Surgery* (Lippincott, Williams & Wilkens, Medford, NJ). He serves on numerous editorial boards, including the *Annals of Surgery, Archives of Surgery, World Journal of Surgery, Journal of the American College of Surgeons*, the *American Journal of Surgery* (Associate Editor), the *Journal of Trauma, Shock, Journal of Surgical Education*, the *American Surgeon*, and others. In addition, he is a reviewer for the *New England Journal of Medicine*. Dr. Britt, a member of Alpha Omega Alpha, is the recipient of the nation's highest teaching award in medicine—the Robert J. Glaser Distinguished Teaching Award, which is given by the Association of American Medical Colleges in conjunction with Alpha Omega Alpha Honor Medical Society. He was honored by the Association of Surgical Education with its lifetime achievement award—the Distinguished Educator Award—given annually to one person considered by his peers to be a true master. More than 200 institutions throughout the world have invited him to be their distinguished visiting professor. He was recently the William P. Longmire, M.D., Visiting Professor at the University of California, Los Angeles. Dr. Britt is the past president of the Society of Surgical Chairs and the past chairman of the ACGME Residency Review Committee for Surgery. In addition, he is the past secretary of the Southern Surgical Association, the past recorder/program chair for the American Association for the Surgery of Trauma, and past president of the Southeastern Surgical Congress, the Halsted Society, and the Southern Surgical Association. Dr. Britt is the past chairman of the Board of Regents of the American

College of Surgeons. He is also the past president of the American College of Surgeons, the American Association for the Surgery of Trauma, and the American Surgical Association. At the inaugural presidential ceremony held in Washington, DC, during the 96th annual Clinical Congress of the American College of Surgeons, Dr. Britt was awarded the U.S. Surgeon General's Medallion for his outstanding achievements in medicine. The Honorable Regina Benjamin, M.D., the 18th U.S. Surgeon General, presented this award at a formal ceremony. Dr. Britt was also appointed to the Robert Wood Johnson Foundation's Clinical Scholar Program National Advisory Committee. The National Library of Medicine of the National Institutes of Health (in collaboration with the Reginald F. Lewis Museum of Maryland African American History and Culture) featured Dr. Britt for his contributions to academic surgery. President George W. Bush recognized Dr. Britt's leadership role in medicine and nominated him to the Board of Regents of the Uniformed Services University (confirmed by the U.S. Senate). At the end of his tenure, Dr. Britt was awarded the coveted Distinguished Service Medal. The National Board of Medical Examiners (NBME) also awarded him the Edithe J. Levit Distinguished Service Award. An active participant in the community, Dr. Britt has received numerous awards for public service. Dr. Britt is the recipient of the 2010 Colgate Darden Citizen of the Year Award and the 2011 Dr. Martin Luther King, Jr. Community Award. The *Atlanta Post* recently highlighted him as one of the top 21 Black doctors in America. *Ebony* magazine recently listed him as one of the most influential African Americans in the nation. At the 2012 annual meeting of the American Surgical Association, Dr. Britt became the 132nd president of the organization. He was conferred an Honorary Doctorate by the president of Tuskegee University. Dr. Britt was also elected to the position of commissioner of the Joint Commission (formerly JACHO). In 2012, he was conferred an Honorary Fellowship in the French Academy of Surgery, and the Colleges of Medicine of South Africa. Having recently been awarded an Honorary Fellowship in the Royal College of Surgeons of Glasgow, Dr. Britt now has the distinction of receiving the highest honor given by each of the four Royal Colleges in the United Kingdom. Dr. Britt, author of the term "Acute Care Surgery" and one of the principal architects of this emerging specialty, was the 2013 recipient of the prestigious Roswell Park Medal for which he was honored for his major contributions to American surgery. At the 2015 annual meeting of the Society of Critical Care Medicine, Dr. Britt was bestowed the unique honor of being recognized as a "Master of Critical Care Medicine" by the American College of Critical

Care Medicine. In 2015, Virginia Governor Terry McAuliffe appointed Dr. Britt to the Board of Visitors of the University of Virginia. In addition, the Southern Surgical Association awarded Dr. Britt the organization's highest accolade—Honorary Fellowship—at its 2015 annual meeting. In the spring of 2016, Dr. Britt was awarded the Urban League of Hampton Roads Professor Marian Capps Memorial Award for his accomplishments in community service through education. At the 148th commencement at Howard University, Dr. Britt was conferred an honorary doctorate (Doctor of Science). His co-honorees included President Barack Obama, activist and award-winning actor Cicely Tyson, and Ambassador Dawson. The summer of 2016 serendipitously, Dr. Britt conducted his 200th visiting professorship at Cook County Hospital in Chicago (where Dr. Britt completed his residency training). He was the inaugural John A. Barrett, M.D., lecturer that kicked off the 50th anniversary celebration for Cook County Hospital Trauma Unit. In 2016, he was elected to the National Academy of Medicine (formerly the Institute of Medicine). He is the first and only faculty member from his institution to receive this distinction—considered one of the highest honors in the field of health and medicine. Dr. Britt was awarded, as the principal investigator, a multimillion dollar NIH-R01 research grant (1 R01 MD011695-01). He was also recently awarded, as the co-investigator (CO-I) another multimillion dollar NIH-RO1 research grant (1 R01 MD011685-01). The Mayo Clinic bestowed upon Dr. Britt one of the institutions highest honors by appointing him to be the Donald Balfour Visiting Professor. In addition to the unveiling of his latest edition of the textbook, *Acute Care Surgery* (Wolters Kluwer), Dr. Britt was inducted into the inaugural class of the Academy of Master Surgeons Educators of the American College of Surgeons in October 2018. He is a founding member of the Academy. In 2019, Dr. Britt was awarded the Silbergleit Award given by the Association of Program Directors in Surgery. He is only the third individual to receive this recognition for his sustained leadership role as an accomplished program director.

Kimberly Bryant is the founder and CEO of Black Girls CODE, a nonprofit organization dedicated to "changing the face of technology" by introducing girls of color (ages 7-17) to the field of technology and computer science with a concentration on entrepreneurial concepts. Ms. Bryant was awarded the Jefferson Award for Community Service for her work to support communities in the Bay Area, named by Business Insider on its list of "The 25 Most Influential African-Americans in Technology," and named to The Root 100 and the Ebony Power 100 lists in 2013. Ms. Bryant was named a White House Champion of Change for her work in tech inclusion and for her focus on bridging the digital divide for girls of color and received an Ingenuity Award in Social Progress from the Smithsonian Institute.

André L. Churchwell is Levi Watkins Jr., M.D., chair and Vanderbilt University Medical Center's chief diversity officer, and a professor of medicine (cardiology), professor of radiology and radiological sciences, professor of biomedical engineering, and senior associate dean for diversity affairs at Vanderbilt University School of Medicine. He was named the 2005 Walter R. Murray Jr. Distinguished Alumnus by the Association of Vanderbilt Black Alumni. The award recognizes lifetime achievements in personal, professional, and community arenas. Dr. Churchwell graduated from the Vanderbilt School of Engineering magna cum laude in 1975. He won the Biomedical Engineering Student Program Award that same year. He received his medical degree from Harvard Medical School in 1979 and later completed his internship, residency, and cardiology fellowship at Emory University School of Medicine and affiliated hospitals in Atlanta. In addition, he was the first African American chief medical resident at Grady Memorial Hospital (1984-1985). Dr. Churchwell received the J. Willis Hurst Award for Best Clinical Teacher in 1991 from Emory, and in 2004 he was named the Emory University School of Medicine Resident Alumni Distinguished Achievement Award winner. For the past 10 years he has been named one of the nation's top cardiologists in "The Best Doctors in America." In 1986, while at Emory University, he was also named Most Outstanding House Officer, made an honorary Morehouse Medical School class member, and he received a Robert Wood Johnson Foundation Minority Medical Faculty Development Award. In 2010, he was awarded the Distinguished Alumnus Award of Vanderbilt University School of Engineering. Along with his physician brothers Kevin and Keith, he received the 2011 Trumpet Award for Medicine. He serves on many medical school committees including the Admission and Promotion Committees and recently was named dean of diversity for undergraduate medical education to add to his current role in the dean's office. In 2012 and 2013, the Vanderbilt University Organization of Black Graduate and Professional Students honored Dr. Churchwell with one of the organization's first Distinguished Faculty Awards. He was also recognized with an American Registry Most Compassionate Doctor Award. From 2010 to 2013, he has been awarded the Professional Research Consultants' Five-Star Excellence Award—Top

10% Nationally for "Excellent" Responses for Medical Specialty Services and Overall Quality. In 2014, he was honored as one of the Top 15 Most Influential African American Medical Educators by *Black Health* magazine. Furthermore, he was elected in 2012 to serve as the southern representative for the Group on Diversity and Inclusion for the AAMC (American Association of Medical Colleges). Since 2011, he has served on the editorial board of the *Cardiovascular Engineering and Technology: A Journal of the Biomedical Engineering Society*. In 2013, he helped create the Hurst-Logue-Wenger Cardiovascular Fellows Society (HLWCFS) of Emory University School of Medicine and was elected the first president of HLWCFS.

Theodore Corbin is vice chair for research and an associate professor in the Department of Emergency Medicine at the Drexel University College of Medicine. He also serves as the medical director of the "Healing Hurt People" Program, an emergency department–based, trauma-informed intervention strategy that identifies victims of intentional injury. Dr. Corbin received his Master of Public Policy from the Woodrow Wilson School at Princeton University. Recipient of the 2017 Drexel University College of Medicine Distinguished Alumni, Dr. Corbin also co-directs the Center for Nonviolence and Social Justice at Drexel University School of Public Health, where he holds a joint appointment. He was awarded a Stoneleigh Foundation Fellowship and an Annie E. Casey Foundation grant to explore the impact of PTSD on violently injured youth and young adults, and to evaluate the effectiveness of Healing Hurt People. His work focuses broadly on addressing the trauma in the lives of victims of violence, especially boys and men of color for whom violence is a leading cause of disability and death.

George Q. Daley, M.D., Ph.D., is Dean and Caroline Shields Walker Professor of Medicine at Harvard Medical School. He is also professor of biological chemistry and molecular pharmacology. Prior to becoming dean he was the director of the Pediatric Stem Cell Transplantation Program at Dana-Farber/Boston Children's Cancer and Blood Disorders Center and an investigator of the Howard Hughes Medical Institute. Dr. Daley received his A.B., magna cum laude, from Harvard (1982), a Ph.D. in biology from the Massachusetts Institute of Technology (1989), working with Nobel Laureate David Baltimore, and his M.D., summa cum laude, from Harvard Medical School (1991). Dr. Daley pursued clinical training in internal medicine at Massachusetts General Hospital, where he served as chief resident

(1994-1995), and a clinical fellowship in hematology/oncology at Brigham and Women's Hospital and Children's Hospitals. He remains a staff member in Pediatric Hematology/Oncology at Boston Children's Hospital. Dr. Daley's research uses mouse and human disease models to study cancer and blood disorders. Dr. Daley has been elected to the National Academy of Medicine, the American Society for Clinical Investigation, the American Association of Physicians, the American Pediatric Societies, the American Academy of Arts and Sciences, and the American Association for the Advancement of Science. He was an inaugural winner of the National Institutes of Health Director's Pioneer Award (2004), and has won the E. Donnall Thomas Prize of the American Society of Hematology. He was a founding executive committee member of the Harvard Stem Cell Institute, served as president of the International Society for Stem Cell Research (2007-2008), and anchored the special task forces that produced the society's guidelines for stem cell research and clinical translation (2006, 2008, 2016). He was on the organizing committee for both the 2015 and 2018 International Summits on Human Genome Editing and has advocated publicly for responsible international guidelines for attempts at germline genome editing.

Wayne A. I. Frederick, M.D., was appointed the 17th president of Howard University in 2014. He previously served as Provost and Chief Academic Officer. A distinguished scholar and administrator, Dr. Frederick has advanced Howard University's commitment to student opportunity, academic innovation, public service, and fiscal stability. Under his leadership, Howard University is now ranked as a Tier 1 national university by *U.S. News & World Report*. Early in his tenure as president, Dr. Frederick pursued initiatives to streamline and strengthen university operations. He has overseen a series of reform efforts, including the expansion of academic offerings, establishing innovative programs to support student success, and the modernization of university facilities.

As an undergraduate, Dr. Frederick was admitted to Howard University's B.S./M.D. dual degree program. He completed the requirements for both degrees in 6 years, allowing him to earn his Bachelor of Science degree and his medical degree by the age of 22. He also earned a Master of Business Administration from Howard University's School of Business in 2011.

Following his post-doctoral research and surgical oncology fellowships at the University of Texas MD Anderson Cancer Center, Dr. Frederick began his academic career as associate director of the Cancer Center at the University of Connecticut. Upon his return to Howard University, his academic positions included associate dean in the College of Medicine, division chief in the Department of Surgery, director of the Cancer Center, and deputy provost for health sciences.

Dr. Frederick is the author of numerous peer-reviewed articles, book chapters, abstracts, and editorials and is a widely recognized expert on disparities in health care and medical education. He continues to operate and also gives lectures to second-year medical students and surgical residents of Howard University's College of Medicine. His medical research focuses on narrowing racial, ethnic, and gender disparities in cancer-care outcomes, especially about gastrointestinal cancers. Dr. Frederick also devotes his time to writing and speaking on salient topics in higher education including the impact of Historically Black Colleges and Universities, campus intellectual diversity, the underrepresentation of African American men in medical school, and gender equity on college campuses.

In 2019, Dr. Frederick was honored with the Distinguished Alumnus Award from the University of Texas MD Anderson Cancer Center for his contributions to the medical field. Dr. Frederick has received various awards honoring his scholarship and service. In January 2017, the Federal Reserve System Board of Governors elected Dr. Frederick to the Federal Reserve Bank of Richmond's Baltimore Branch. He was presented with the Diaspora Public Diplomacy Leadership Award by the Embassy of the Republic of Trinidad and Tobago for his contributions to strengthening Trinidad and Tobago-United States bilateral relations through excellence in global educational leadership. In May 2016, President Barack H. Obama appointed Dr. Frederick to the Board of Advisors for the White House Initiative on Historically Black Colleges and Universities. Dr. Frederick has also received the National Association of Health Services Executives' Congressional Black Caucus Distinguished Leadership in Health Care Award, and a Congressional Citation for Distinguished Service, presented by the Honorable Barbara Lee on the Occasion of Caribbean-American Heritage Month. In 2015, Dr. Frederick was also recognized by the then president of the Republic of Trinidad and Tobago for his appointment as president of Howard University. Most recently, Dr. Frederick was appointed to the Board of Directors of the U.S. Chamber of Commerce. Dr. Frederick is a member of surgical and medical associations, including the American Surgical Association and the American College of Surgeons.

Dr. Frederick has also been featured as one of "America's Best Physicians" by *Black Enterprise* magazine. He was named one of *Ebony* magazine's "Power 100," and recognized as a "Super Doctor" in *The Washington Post Magazine*. In 2017, he was named "Washingtonian of the Year" by *Washingtonian* magazine and in 2015 was named "Male President of the Year" by *HBCU Digest* and was inducted into the St. Mary's College, Port of Spain, Trinidad Hall of Fame.

Garth Graham is a leading authority on social determinants of health. President of the Aetna Foundation and vice president of community health for Aetna, Inc., he is also a cardiologist and public health expert. Dr. Graham oversees the community health initiatives for the Foundation and Aetna, Inc., bringing his experience as a former deputy assistant secretary at the U.S. Department of Health and Human Services under the Obama and Bush administrations, where he also ran the Office of Minority Health. He directed the development of the federal government's first National Health Disparities Plan released under the Obama administration. Dr. Graham has been a contributor to *The Hill*, the *Chicago Tribune*, *Fortune*, *Quartz*, *Health Affairs*, and *Ebony*, and has been featured in *Essence*, *CNN*, and the *New York Times* among others. His original research has been published in the *Journal of the American Medical Association*, *American Journal of Public Health*, *Health Affairs*, and other publications. Along with his role at the Aetna Foundation, Dr. Graham is a clinical associate professor of medicine at the University of Connecticut. Prior to joining the Foundation, in his role as the assistant dean for health policy at the University of Florida School of Medicine, Dr. Graham led several research initiatives looking at how to improve outcomes and readmission rates in cardiac patients in underserved populations. He contributes to several boards including being named by the President to the U.S. Federal Coordinating Council on Comparative Effectiveness Research, the Institute of Medicine Board on Population Health, the American Heart Association/American Stroke Associational National Quality Oversight Committee, the American College of Cardiology/American Heart Association Task Force on Clinical Data Standards and many others. Dr. Graham holds a medical degree from Yale School of Medicine, a master's in public health from Yale School of Public Health, and a bachelor of science in biology from Florida International University in Miami. He completed clinical training in cardiology and interventional cardiology at Massachusetts General Hospital and Johns Hopkins. He holds three board certifications: internal medicine, cardiology, and interventional cardiology.

Paula T. Hammond is a David H. Koch Professor in Engineering and the Head of the Department of Chemical Engineering at the Massachusetts Institute of Technology. Her laboratory designs polymers and nanoparticles for drug delivery and energy-related applications including batteries and fuel cells. She is an intramural faculty member of the Koch Institute for Integrative Cancer Research and an Associate Editor of *ACS Nano*.

Evelyn Hammonds is a member of the faculty in the Faculty of Arts and Sciences at Harvard University. She was the first senior vice provost for faculty development and diversity at Harvard (2005-2008). From 2008 to 2013 she served as dean of Harvard College. She holds honorary degrees from Spelman College and Bates College. Professor Hammonds is the director of the Project on Race & Gender in Science & Medicine at the Hutchins Center for African and African American Research at Harvard. Professor Hammonds earned a Ph.D. in the history of science from Harvard University, an S.M. in physics from the Massachusetts Institute of Technology (MIT), a B.E.E. in electrical engineering from the Georgia Institute of Technology, and a B.S. in physics from Spelman College. In 2010 she was appointed to President Barack Obama's Board of Advisers on Historically Black Colleges and Universities and in 2014 to the President's Commission on Excellence in Higher Education for African Americans. She has published articles on the history of disease, race and science, African American feminism, African American women and the epidemic of HIV/AIDS and analyses of gender and race in science and medicine. Professor Hammonds' current research focuses on diversity in STEM fields; the intersection of scientific, medical and socio-political concepts of race in the United States; and genetics and society. Professor Hammonds served two terms on the Committee on Equal Opportunity in Science and Engineering (CEOSE), the congressionally mandated oversight committee of the National Science Foundation (NSF). Professor Hammonds was appointed to the Committee on Women in Science, Engineering, and Medicine (CWSEM) of the National Academies in 2017. She was elected to the National Academy of Medicine (NAM) in 2018.

Ian D. Henry, Ph.D., is a section head in R&D at Procter & Gamble. A native of Marion, IN, Dr. Henry earned his B.A. in chemistry from Earlham College in 2001 and a Ph.D. in analytical chemistry from Purdue University in 2008, where he studied under Dr. M. Daniel Raftery. Currently, Dr. Henry leads the Analytical group for P&G's global feminine care business.

Prior to Feminine Care, Dr. Henry led the Qualitative Mass Spectrometry group in the Trace Analysis Capability and the Analytical Digital Platforms group in corporate R&D. An analytical chemist with a background in bioanalytical NMR Spectroscopy, Dr. Henry started his P&G journey in the Beauty business, supporting innovation programs for brands such as Olay, Safeguard, Pantene, and Head & Shoulders. During his tenure in Beauty, he was an original member of the Centric Team, a grassroots-led group of Black Ph.D. scientists who led fundamental hair studies and value proposition creation that resulted in the startup of focused product initiatives for consumers of African ancestry, most notably Pantene Gold Series, H&S Royal Oils and, more recently, the My Black Is Beautiful brand. The team's work earned both CTO Pathfinder and P&G Diversity and Inclusion Award honors. In 2016, Dr. Henry was selected as a Great Leader Under 40 by LEAD Cincinnati. Beyond work, Dr. Henry is the vice president of the Cincinnati Chapter of NOBCChE and active in the local Cincinnati Section of the ACS, where he is involved in STEM outreach throughout the greater Cincinnati region. Since 2012, Dr. Henry has been a member of the Board of Trustees at Earlham College, where he leads the Diversity Committee. He is also a mentor in the Big Brothers Big Sisters program, serving since 2010.

Lynne M. Holden, M.D. is the co-founder and president of Mentoring in Medicine, Inc. (MIM). MIM is a national health and science youth development nonprofit organization. The mission of MIM is to expose, inspire, educate, and equip students to become biomedical professionals through academic enrichment, leadership development, civic engagement, and mentoring. MIM has reached nearly 52,000 students, parents, and educators from elementary school through medical school and recruited 1,500 health and science volunteers. Dr. Holden provides the overall leadership, creates the organizational strategy, recruits volunteers, facilitates program development, and establishes collaborative partnerships.

Dr. Holden earned her B.S. in zoology from Howard University, graduated from Temple University School of Medicine and completed her residency in emergency medicine at the Jacobi/Montefiore Emergency Medicine Residency Program. She is a practicing emergency department physician at Montefiore Health System in the Bronx, NY. She is a professor of emergency medicine at the Albert Einstein College of Medicine, where she has served as a co-chair of the Admissions Committee and in various leadership positions in the Emergency Medicine Residency Program, the largest in the country. Dr. Holden serves on several national boards including the Friends of the National Library of Medicine and the CUNY School of Medicine. She is active in the National Medical Association on the local, regional, and national levels. She is a deacon at the Abyssinian Baptist Church in Harlem, NY, and a member of Delta Sigma Theta, Inc.

Mentoring in Medicine has earned 60 press features including *JET*, *Essence*, *CNN*, the *New York Times*, and *FOX News*. Dr. Holden has published extensively and received numerous awards for her work, including the Maybelline NY-Essence Empowerment through Education Award (2007), Society of Academic Emergency Visionary Educator Award (2008), Robert Wood Johnson Foundation Community Health Leader (2009), Washington Post Root 100 Leader (2010), Lifetime TV Remarkable Woman (2010), American Medical Association Inspirational Physician (2016), and the United Hospital Fund Distinguished Community Service Award (2019).

Camara Phyllis Jones, M.D., MPH, Ph.D., is the 2019-2020 Evelyn Green Davis Fellow at the Radcliffe Institute for Advanced Study at Harvard University, and the 144th president of the American Public Health Association (2016). She is a family physician and epidemiologist whose work focuses on naming, measuring, and addressing the impacts of racism on the health and well-being of the nation. She seeks to broaden the national health debate to include not only universal access to high-quality health care, but also attention to the social determinants of health (including poverty) and the social determinants of equity (including racism).

Dr. Jones is a public health leader valued for her creativity and intellectual agility. As a methodologist, she has developed new methods for comparing full distributions of data, rather than simply comparing means or proportions, in order to investigate population-level risk factors and propose population-level interventions. As a social epidemiologist, her work on "race"-associated differences in health outcomes goes beyond simply documenting those differences to vigorously investigating the structural causes of the differences. As a teacher, her allegories on "race" and racism illuminate topics that are otherwise difficult for many Americans to understand or discuss. She aims through her work to catalyze a National Campaign Against Racism that will mobilize and engage all Americans.

Dr. Jones was an assistant professor at the Harvard School of Public Health (1994-2000) before being recruited to the U.S. Centers for Disease Control and Prevention (2000-2014), where she served as a medical officer and research director on social determinants of health and equity. Most recently, she was a senior fellow at the Satcher Health Leadership Institute and the Cardiovascular Research Institute at the Morehouse School of Medicine (2013-2019). She has been elected to service on many professional boards, including her current service on the Board of Directors of the DeKalb County (Georgia) Board of Health, and the National Board of Public Health Examiners.

She is also actively sought as a contributor to national efforts to eliminate health disparities and achieve health equity, including as a faculty member of the Accreditation Council for Graduate Medical Education's Pursuing Excellence in the Clinical Learning Environment collaborative addressing Health Care Disparities, as a member of the National Academies of Sciences, Engineering, and Medicine's Roundtable on Black Men and Black Women in Science, Engineering, and Medicine, and as a project advisor and on-screen expert for the groundbreaking film series *Unnatural Causes: Is Inequality Making Us Sick?*

Highly valued as a mentor and teacher, she is also an adjunct professor at the Rollins School of Public Health at Emory University and an adjunct associate professor at the Morehouse School of Medicine. Her honors include the Wellesley Alumnae Achievement Award (Wellesley College's highest alumnae honor, 2018), the John Snow Award (given in recognition of "enduring contributions to public health through epidemiologic methods and practice" by the American Public Health Association's Epidemiology Section, 2011), and awards named after luminaries David Satcher (2003), Hildrus A. Poindexter (2009), Paul Cornely (2016), Shirley Nathan Pulliam (2016), Louis Stokes (2018), Frances Borden-Hubbard (2018), and Cato T. Laurencin (2018).

Dr. Jones earned her B.A. in molecular biology from Wellesley College, her M.D. from the Stanford University School of Medicine, and both her Master of Public Health and her Ph.D. in epidemiology from the Johns Hopkins School of Hygiene and Public Health. She also completed residency training in general preventive medicine at Johns Hopkins and in family practice at the Residency Program in Social Medicine at Montefiore Medical Center.

Orlando C. Kirton is chairman of surgery and surgeon-in-chief at Abington Jefferson Health. Dr. Kirton received his undergraduate degree from Brown University in Providence, RI, and his medical degree, cum laude, from Harvard Medical School. He served his internship and residency in surgery at SUNY and then completed fellowships in surgical critical care and surgery of trauma at Jackson Memorial Hospital, Department of Surgery and University of Miami School of Medicine in Florida. He joined the faculty at the University of Miami from 1992 to 1999 where he achieved the academic rank of associate professor of surgery and served as the director of the Trauma Intensive Care Unit at Jackson Memorial Hospital and served as the interim director of the Trauma Service. From 1999 to 2016 Dr. Kirton was the *Ludwig J. Pyrtek, MD Chair in Surgery*, chief of the Department of Surgery, chief of the Division of General Surgery, and associate director of the Surgical Intensive Care Unit. He also was chief of trauma at Hartford Hospital from 2012-2016. Dr. Kirton's current academic rank is that of professor of surgery and vice-chairman of the Department of Surgery of the Sidney Kimmel Medical College at Thomas Jefferson University. Dr. Kirton is a diplomat of the American Board of Surgery with additional qualification in Surgical Critical Care. He is a fellow of the American College of Surgeons, the American College of Critical Care Medicine, and the American College of Chest Physicians, and member of the Society of University Surgeons and the American Surgical Association. In 2014 the Society of Critical Care medicine bestowed him the Master of Critical Care medicine distinction. Dr. Kirton has served as president of the Surgical Section of the National Medical Association, the president of the Society of Black Academic Surgeons, and was also past president of the Connecticut Chapter of the American College of Surgeons. He served on the Boards of Directors for the Society of Critical Care Medicine, the Eastern Association for the Surgery of Trauma, the National Medical Association, the Society of Black Academic Surgeons, and the Board of Managers of The American Association for the Surgery of Trauma. Dr. Kirton Received a Physician Executive M.B.A. from the University of Tennessee in 2015. He has published extensively in peer-reviewed, referred journals as well as authored numerous chapters and textbooks on surgical critical care, trauma, and surgical education.

John R. Lumpkin, M.D., M.P.H., is president of the Blue Cross and Blue Shield of North Carolina Foundation, a position he has held since April 2019. The Foundation seeks to improve the health and well-being of all North Carolinians through a focus on transforming the health care system (including oral health), expanding access to healthy food, supporting a healthy start in life for children, improving the physical conditions where people live, and strengthening the ability of communities to improve health.

Dr. Lumpkin most recently served as senior vice president, Programs for the Robert Wood Johnson Foundation (RWJF). At RWJF, Dr. Lumpkin was responsible for the Foundation's efforts aimed at transforming health and health care systems, ensuring that everyone has access to stable and affordable health care coverage, building leadership, and engaging business toward building a Culture of Health in the United States. These efforts helped to catalyze fundamental changes in health and health care systems to achieve measurably better outcomes for all by maintaining high-quality, effective, and value-laden health care, public health, and population health services.

Before joining RWJF in April 2003, Dr. Lumpkin served as director of the Illinois Department of Public Health for 12 years. During his more than 17 years with the department, he served as acting director and prior to that as associate director.

Dr. Lumpkin has participated directly in the health and health care system, first practicing emergency medicine and teaching medical students and residents at the University of Chicago and Northwestern University. He is the past chairman of the board of directors of the Robert Wood Johnson University Hospital, the major teaching hospital of Rutgers University in New Brunswick. After earning his M.P.H. in 1985, he began caring for the more than 12 million people of Illinois as the first African American director of the state public health agency with more than 1,300 employees in seven regional offices, three laboratories, and locations in Springfield and Chicago. He led improvements to programs dealing with women's and men's health, information and technology, emergency and bioterrorism preparedness, infectious disease prevention and control, immunization, local health department coverage, and the state's laboratory services.

Dr. Lumpkin is a member of the National Academy of Medicine and a fellow of the American Academy of Nursing, American College of Emergency Physicians and the American College of Medical Informatics. He has been chairman of the National Committee on Vital and Health Statistics, and served on the U.S. Department of Agriculture's Council on Maternal, Infant, and Fetal Nutrition, the advisory committee to the director of the U.S. Centers for Disease Control and Prevention, and the National Institute of Medicine's Committee on Assuring the Health of the Public in the 21st Century. He has served on the boards of directors for the Public Health Foundation and National Quality Forum, as president of the Illinois College of Emergency Physicians and the Society of Teachers of Emergency Medicine, and as speaker and board of director's member of the American College of Emergency Physicians. He has received the Arthur McCormack Excellence and Dedication in Public Health Award from the Association of State and Territorial Health Officials (ASTHO), the Jonas Salk Health

Leadership Award, and the Leadership in Public Health Award from the Illinois Public Health Association. Dr. Lumpkin also has been the recipient of the Bill B. Smiley Award, Alan Donaldson Award, African American History Maker, and Public Health Worker of the Year of the Illinois Public Health Association. He is the author of numerous journal articles and book chapters.

Dr. Lumpkin earned his M.D. and B.M.S. degrees from Northwestern University Medical School and his M.P.H. from the University of Illinois School of Public Health. He was the first African American trained in emergency medicine in the country after completing his residency at the University of Chicago. He has served on the faculty of the University of Chicago, Northwestern University, and University of Illinois at Chicago.

Shirley Malcom is Head of Education and Human Resources Programs of the American Association for the Advancement of Science (AAAS). The directorate includes AAAS programs in education, activities for underrepresented groups, and public understanding of science and technology. Dr. Malcom was head of the AAAS Office of Opportunities in Science from 1979 to 1989. Between 1977 and 1979, she served as program officer in the Science Education Directorate of the National Science Foundation (NSF). Prior to this, she held the rank of assistant professor of biology, University of North Carolina, Wilmington, and for 2 years was a high school science teacher. Dr. Malcom serves on several boards, including the Howard Heinz Endowment. She is an honorary trustee of the American Museum of Natural History, a Regent of Morgan State University, and a trustee of Caltech. She has chaired a number of national committees addressing education reform and access to scientific and technical education, careers and literacy. Dr. Malcom is a former trustee of the Carnegie Corporation of New York and a fellow of the AAAS and the American Academy of Arts and Sciences. In 2003, she received the Public Welfare Medal of the National Academy of Sciences, the highest award bestowed by the Academy.

Cora Bagley Marrett is the former deputy director of the National Science Foundation (NSF), a position she held from 2011 to 2014. She previously held the position of senior advisor (2009-2011), except for 6 months when she served as the Foundation's acting director. She has also been a professor of sociology at the University of Wisconsin, where she has held tenure since 1974; she took leave from the university in 2007 to join the NSF as assistant director for education and human resources. From 1992 to 1996 Dr.

Marrett was assistant director for social, behavioral, and economic sciences at the NSF. From 1996 to 1998 she served by appointment on the Board of Governors of the Argonne National Laboratory and was a member of a peer-review oversight group for the National Institutes of Health. From 1997 to 2001, she was provost, senior vice chancellor for academic affairs, and a professor of sociology and Afro-American studies at the University of Massachusetts-Amherst. Throughout her career, Dr. Marrett has worked to expand opportunities for minorities. She is credited with having brought scholars of color into the field of sociology and with working actively to improve conditions of inequality revealed by sociological research. Dr. Marrett earned a B.A. in sociology from Virginia Union University and M.A. and Ph.D. degrees, also in sociology, from the University of Wisconsin-Madison. She received an honorary doctorate from Wake Forest University in 1996 and from Virginia Union University in 2011. She was elected a fellow of the American Academy of Arts and Sciences in 1996 and of the American Association for the Advancement of Science in 1998. In 2008 the American Sociological Association recognized her many contributions with the Johnson-Cox-Frazier Award, and the Wisconsin Alumni Association honored her with its Distinguished Alumni Award in 2012. She is a member of the Board of Visitors of the University of Wisconsin-Madison College of Letters and Science.

Alfred Mays is a program officer at the Burroughs Wellcome Fund. Mr. Mays is responsible for managing grant competitions in science education and diversity of science. He also works closely with the North Carolina (NC) Science, Mathematics, and Technology Education Center. Prior to Mr. Mays assuming this role, he served as an independent consultant with a service delivery that included strategic planning, project incubation, design, and implementation of a number of initiatives within education agencies and organizations. Mr. Mays was the founder of EdSync Strategies, Inc., an education contract service that provided assistance to NC eLearning Commission, NC STEM (science, technology, engineering, and mathematics) Learning Network, rural NC public school systems, and the Public School Forum. From 2007 to 2011, he served as the assistant director of the Collaborative Project, an initiative that "sought to strengthen participating school systems serving low-income students in rural areas of the state." Mr. Mays has also worked with the University of North Carolina General Administration, serving as the director of information resources and director of special projects. Mr. Mays received his B.S. from Wilmington

College and his M.S. in administration from Central Michigan University. He served in the U.S. Air Force from 1984 to 1994, providing information system and data management support for various U.S. Air Force missions.

Valerie Montgomery Rice, M.D., FACOG, provides a valuable combination of experience at the highest levels of patient care and medical research, as well as organizational management and public health policy. Marrying her transformational leadership acumen and strategic thinking to tackle challenging management issues, she has a track record of redesigning complex organizations' infrastructures to reflect the needs of evolving strategic environments and position the organization for success through sustainability tactics.

The sixth president of Morehouse School of Medicine (MSM) and the first woman to lead the freestanding medical institution, Dr. Montgomery Rice serves as both the president and dean. A renowned infertility specialist and researcher, she most recently served as dean and executive vice president of MSM, where she has served since 2011.

Prior to joining MSM, Dr. Montgomery Rice held faculty positions and leadership roles at various health centers, including academic health centers. Most notably, she was the founding director of the Center for Women's Health Research at Meharry Medical College, one of the nation's first research centers devoted to studying diseases that disproportionately impact women of color.

Dedicated to the creation and advancement of health equity, Dr. Montgomery Rice lends her vast experience and talents to programs that enhance pipeline opportunities for academically diverse learners, diversifies the physician and scientific workforce, and fosters equity in health care access and health outcomes. She holds memberships in various organizations and participates on a number of boards, such as the following: member, National Academy of Medicine, and board of directors for National Center for Advancing Translational Sciences, The Metro Atlanta Chamber, Kaiser Permanente School of Medicine, The Nemours Foundation, UnitedHealth Group, Westside Future Fund, Josiah Macy Jr. Foundation, the Association of American Medical Colleges Council of Deans, and Horatio Alger Association.

Dr. Montgomery Rice has received numerous accolades and honors. She was named to the Horatio Alger Association of Distinguished Americans and received the 2017 Horatio Alger Award. For three consecutive years (2016-2018) *Georgia Trend Magazine* selected Dr. Montgomery Rice as one of the 100 Most Influential Georgians. Other honors include the following: The Turknett Leadership Character Award (2018), Visions of Excellence Award, Atlanta Business League (2018), Links Incorporated Co-Founders Award (2018), Trumpet Vanguard Award (2015), The Dorothy I. Height Crystal Stair Award (2014), National Coalition of 100 Black Women—Women of Impact (2014), YWCA—Women of Achievement of Atlanta (2014) and Nashville (2007), American Medical Women's Association Elizabeth Blackwell Medal (2011), and Working Mother Media Multicultural Women's Legacy Award (2011).

A Georgia native, Dr. Montgomery Rice holds a bachelor's degree in chemistry from the Georgia Institute of Technology, a medical degree from Harvard Medical School, and an honorary degree from the University of Massachusetts Medical School and Doctor of Humane Letters honorary degree from Rush University; all reflect her lifetime commitment to education, service, and the advancement of health equity. She completed her residency in obstetrics and gynecology at Emory University School of Medicine and her fellowship in reproductive endocrinology and infertility at Hutzel Hospital.

Randall C. Morgan is the executive director of the W. Montague Cobb/NMA Health Institute and an orthopedic surgeon who practices in Sarasota and Bradenton, Florida. He serves as founder and president of University Park Orthopedics in that community. He is also clinical associate professor of orthopedic surgery at Florida State University School of Medicine. Dr. Morgan also served as the 95th president of the National Medical Association during the years 1996 and 1997. He was the first board certified orthopedic surgeon to hold that position. Dr. Morgan is a true pioneer in his profession and was among the first surgeons to perform total joint replacement surgery at Northwestern University. Dr. Morgan has practiced medicine in Evanston, IL, as well as in his hometown of Gary, IN, for more than 30 years prior to his relocation to Sarasota. With the assistance of his father, Mr. Randall C. Morgan, Sr., he founded the Orthopedic Centers of Northwest Indiana and served as its president from 1975 to 1999. The center was once the largest minority-owned orthopedic practice in the United States. He is a diplomat of the American Board of Orthopedic Surgery and the American Board of Managed Care Medicine. He is also a fellow of the American College of Surgeons.

Elizabeth Ofili is professor of medicine in cardiology at the Morehouse School of Medicine and chief medical officer of the Morehouse Choice

Accountable Care Organization. She is a national and internationally recognized clinician scientist with particular focus on cardiovascular disparities and women's health. Dr. Ofili has been continuously funded by the NIH and industry/foundations since 1994, with a track record in clinical trials that impact health disparities. In 2002, as president of the Association of Black Cardiologists, she led the initiative to implement the landmark African American Heart Failure Trial, whose findings led to a change in practice guidelines for the treatment of heart failure in African Americans. Over the past 17 years, she has led the growth of the clinical research infrastructure and training programs at Morehouse School of Medicine with awards totaling more than $150 million, including serving as the founding director of the U54 center of clinical research excellence, the community physicians network, the U54 Research Centers in Minority Institutions (RCMI) Center of Excellence for Clinical and Translational Research, and the R25 clinical research education and career development program. Dr. Ofili has mentored more than 30 M.D. and Ph.D. clinical and translational science investigators, many of who remain at MSM. She has mentored more than 25 underrepresented minority STEM undergraduates and high school students through funding from NASA and the Minority Biomedical Research Students program. She is the senior co-PI of the Atlanta Clinical and Translational Science Institute (ACTSI), a citywide collaborative Clinical and Translational Science Awards program at Emory University, Morehouse School of Medicine, and Georgia Institute of Technology, along with their partnering health systems and statewide research organizations. Since 2007, ACTSI has engaged more than 673 investigators and 134 postdoctoral and predoctoral trainees in discovery science, training, and community engagement. Dr. Ofili has led successful multi-institutional collaborations through the ACTSI and the RCMI Translational Research Network of 18 historically Black, Hispanic, and Minority Serving Institutions (MSIs) across the nation, and was lead author of a publication on models of partnerships between HBCUs/MSIs and research intensive institutions. Dr. Ofili holds a patent for "a system and method for chronic illness care," and is the recipient of more than 20 national and international awards, including the 2003 National Library of Medicine's "Changing the Face of Medicine, the Rise of America's Women," the Daniel Savage Memorial Science Award from the Association of Black Cardiologists, America's Top Doctors by *Black Enterprise* magazine and 100 Most Influential Health Care Leaders by *Atlanta Business Chronicle*. She has delivered more than 600 scientific presentations and published more than 130 scientific papers in national and international journals. As an

AAMC 2007 Council of Dean Fellow, Dr. Ofili led a project on best practices to sustaining the biomedical and physician workforce. She has advised the NIH on diversity in the biomedical research workforce, and currently serves on the Advisory Board of the National Clinical Center (NIH), and on the AAMC advisory panel on research. She is an elected member of the Association of University Cardiologists, and is on the board of directors of the National Space Biomedical Research Institute.

Vivian Pinn was the first full-time director of the National Institutes of Health (NIH) Office of Research on Women's Health, an appointment she held since 1991 and as NIH associate director for research on women's health since 1994 prior to her retirement in August 2011. Since her retirement, she has been named as a senior scientist emerita at the NIH Fogarty International Center. Dr. Pinn came to NIH from Howard University College of Medicine in Washington, DC, where she had been professor and chair of the Department of Pathology from 1982 until 1991. Dr. Pinn had previously held teaching appointments at Harvard Medical School and Tufts University, where she was also assistant dean for student affairs. A special tribute by Senator Olympia Snowe on Dr. Pinn's retirement was published in the Congressional Record in November 2011 commending her contributions during her NIH tenure. The Association of American Medical Colleges awarded her a Special Recognition Award for exceptional leadership over a 40-year career. She has received numerous honors and recognitions, and is a fellow of the American Academy of Arts and Sciences and was elected to the National Academy of Medicine (formerly the Institute of Medicine) in 1995. A graduate and Alumna Achievement Award recipient as well as former Trustee of Wellesley College, she earned her M.D. from the University of Virginia School of Medicine, the only woman or minority in her class. She completed her postgraduate training in Pathology at the Massachusetts General Hospital. Dr. Pinn has received 17 Honorary Degrees of Science, Law, and Medicine, and the University of Virginia School of Medicine has named one of its four advisory medical student colleges as "The Pinn College" in her honor. Tufts University School of Medicine in 2011 announced the "The Vivian W. Pinn Office of Student Affairs," named for her at the time her former medical students dedicated a scholarship in her name. She has held leadership positions in many professional organizations, including President of the National Medical Association (NMA) and is currently Chair of the NMA Past Presidents Council. Dr. Pinn currently serves on the Board of Trustees/Advisors of

Thomas Jefferson University and Tufts University School of Medicine. She was recently elected to Modern Healthcare's Hall of Fame, the first African American woman to be so honored, and received the Outstanding Woman Leader in Healthcare Award from the University of Michigan. Dr. Pinn also holds the position of Professor, Institute for Advanced Discovery and Innovation at the University of South Florida.

Joan Y. Reede is the dean for diversity and community partnership and professor of medicine at Harvard Medical School (HMS). Dr. Reede also holds appointments as professor in the Department of Social and Behavioral Sciences at the Harvard T.H. Chan School of Public Health, and is an assistant in health policy at Massachusetts General Hospital. Dr. Reede is responsible for the development and management of a comprehensive program that provides leadership, guidance, and support to promote the increased recruitment, retention, and advancement of underrepresented minority, women, LGBT, and faculty with disabilities at HMS. This charge includes oversight of all diversity activities at HMS as they relate to faculty, trainees, students, and staff. Dr. Reede also serves as the director of the Minority Faculty Development Program; program director of the Faculty Diversity Program of the Harvard Catalyst/The Harvard Clinical and Translational Science Center, and chair of the HMS Task Force on Diversity and Inclusion. Dr. Reede has served on a number of boards and committees including the Secretary's Advisory Committee to the Director of the National Institutes of Health; the Sullivan Commission on Diversity in the Healthcare Workforce; the National Children's Study Advisory Committee of the Eunice Kennedy Shriver National Institute of Child Health and Human Development, and the Advisory Committee to the Deputy Director for Intramural Research of the National Institutes of Health. Some of her past affiliations include the Steering Committee and Task Force for the Annual Biomedical Research Conference for Minority Students (ABRCMS); past co-chair of the Bias Review Committee of the Advisory Committee to the NIH Director's Working Group on Diversity; the Association of American Medical Colleges Careers in Medicine Committee (AAMC); past chair of the AAMC Group on Diversity and Inclusion (GDI). Dr. Reede served on the editorial board of the *American Journal of Public Health*, and she was the guest editor for the AAMC 2012 special issue, *"Diversity and Inclusion in Academic Medicine"* of *Academic Medicine*. She is a past chair of the National Academy of Medicine's Interest Group 08 on Health of Populations/Health Disparities. In 2018, Dr. Reede was appointed to the National Advisory

Council on Minority Health and Health Disparities (NACMHD). Dr. Reede is an authority in the area of workforce development and diversity. Her colleagues and mentees have recognized her with a number of awards that include the Herbert W. Nickens Award from AAMC and the Society of General Medicine in 2005; election to the National Academy of Medicine in 2009; the 2011 Diversity Award from the Association of University Professors; and in 2012 she was the recipient of an Elizabeth Hurlock Beckman Trust Award. In 2013 she received an Exemplar STEM Award from the Urban Education Institute at North Carolina A & T University in Greensboro, North Carolina, and in 2015, she was the Distinguished Woman Scientist and Scholar ADVANCE Lecturer at the University of Maryland School of Public Health. Dr. Reede was recognized by her medical school classmates as a recipient of The Mount Sinai Alumni Association and Icahn School of Medicine 2015 Jacobi Medallion for extraordinary leaders in health care, and in 2017 she was nominated by her peers, and received a Harvard T.H. Chan School of Public Health Alumni Award.

Louis W. Sullivan is the chairman and chief executive officer of The Sullivan Alliance to Transform the Health Professions. He is also chairman of the board of the National Health Museum in Atlanta, Georgia, which aims to improve the health of Americans by enhancing health literacy and advancing healthy behaviors. Dr. Sullivan served as chair of the President's Commission on Historically Black Colleges and Universities from 2002–2009 and was cochair of the President's Commission on HIV and AIDS from 2001 to 2006. With the exception of his tenure as secretary of the U.S. Department of Health and Human Services (HHS) from 1989 to 1993, Dr. Sullivan was president of Morehouse School of Medicine (MSM) for more than 2 decades. As Secretary of HHS, Dr. Sullivan's efforts to improve the health and health behavior of Americans included (1) the introduction of a new and improved FDA food label; (2) release of Healthy People 2000, a guide for improved health promotion/disease prevention activities; (3) education of the public about health dangers from tobacco use; (4) successful efforts to prevent the introduction of "Uptown," a non-filtered, mentholated cigarette by R.J. Reynolds Tobacco Company; (5) inauguration of a $100 million minority male health and injury prevention initiative; and (6) implementation of greater gender and ethnic diversity in senior positions of HHS, including the appointment of the first female director of the NIH, the first female and first Hispanic Surgeon General of the U.S. Public Health Service, and the first African American Commissioner of the Social Security Administration.

Lamont Terrell graduated salutatorian from Texas Southern University (TSU) as a Frederick Douglas honor scholar earning a B.S degree in chemistry in 1995. While at TSU, his life as a research scientist began doing undergraduate research focusing on the synthesis of inorganic compounds with environmental applications. He earned his Ph.D. in 2001 in organic chemistry from Michigan State University (MSU) under the direction and guidance of Professor Robert Maleczka. His graduate studies consisted of the total synthesis of the antiluekemic natural product amphidinolide A and the development of catalytic tin hydride reactions. Upon completion of his graduate studies at MSU, he continued his synthetic training with a 2-year postdoctoral stint with Professor Barry Trost at Stanford University. The focus of his postdoctoral studies was the development of a catalytic dinuclear zinc asymmetric Mannich reaction. He began his career in drug discovery as a medicinal chemist at GlaxoSmithKline (GSK) in 2003 in its cardiovascular medicinal chemistry group. He spent 11 years doing small molecule lead optimization primarily focusing on cardiovascular disease targets. Outside of leading science, Dr. Terrell is passionately involved with community and outreach efforts. He has been involved with the recruitment of scientists at all levels into the chemistry community. He leads the GSK chemistry recruitment team for minority conferences and serves as the lead for the African American Alliance employee resource group in the Delaware Valley. He is a leader in GSK's inclusion and diversity efforts and a member of the R&D Inclusion council. In 2017, he decided to step away from doing science to lead the U.S. R&D Early Talent Programs and head GSK's diversity recruitment initiative for the U.S. Pharma R&D business.

Hannah Valantine received her M.B.B.S. degree (Bachelor of Medicine, Bachelor of Surgery—the United Kingdom's equivalent to an M.D.) from St. George's Hospital, London University in 1978. After that, she moved to the University of Hong Kong Medical School for specialty training in elective surgery before returning to the U.K. She was awarded a diploma of membership by the Royal College of Physicians (M.R.C.P.) in 1981. In addition, she completed postgraduate training and numerous fellowships, serving as senior house officer in Cardiology at Brompton Hospital and Registrar in Cardiology and General Medicine at Hammersmith Hospital. In 1985, Dr. Valantine moved to the United States for postdoctoral training in cardiology at Stanford University, and in 1988, she received a Doctor of Science (DSc), Medicine, from London University. Dr. Valantine became a Clinical Assistant Professor in the Cardiology Division at Stanford and rose through the academic ranks to become a full professor of medicine in the Division of Cardiovascular Medicine and director of heart transplantation research. She came to the NHLBI in 2014 to continue her research while also serving as the first NIH chief officer of scientific workforce diversity. Dr. Valantine has received numerous awards throughout her career including a Best Doctor in America honor in 2002. She has authored more than 160 primary research articles and reviews and previously served on the editorial boards of the journals *Graft* and *Ethnicity & Disease*. Dr. Valantine is a member of the American College of Cardiology, the American Society of Transplant Physicians, and the American Heart Association, and past President of the American Heart Association Western States Affiliate.

Clyde W. Yancy is a cardiologist at Northwestern Memorial Hospital, the chief of Cardiology Medicine, and the Magerstadt Professor of Medicine and Northwestern University Feinberg School of Medicine. Dr. Yancy has received recognition for clinical and research expertise in the field of heart failure and has additional interests in cardiomyopathy, heart valve diseases, hypertension and prevention. He is an active member of the American Heart Association, American College of Cardiology, American College of Physicians, and the Heart Failure Society of America. His bibliography includes more than 250 peer-reviewed manuscripts, numerous book chapters, editorials, and review articles, consultations for the FDA, NIH, Agency for Healthcare Research and Quality (AHRQ), and Patient-Centered Outcomes Research Institute (PCORI). He has also received numerous Best Physician and Best Teaching Awards.

BIOGRAPHICAL SKETCHES OF WORKSHOP PRESENTERS

David A. Acosta, M.D., provides strategic vision and leadership for the Association of American Medical College's diversity and inclusion activities across the medical education community, and leads the association's Diversity Policy and Programs unit. Dr. Acosta, a family medicine physician, joined the AAMC from the University of California (UC), Davis School of Medicine, where he served as senior associate dean for equity, diversity, and inclusion and associate vice chancellor for diversity and inclusion, and chief diversity officer for UC Davis Health System. He previously served as the first chief diversity officer at the University of Washington School of Medicine, where he established the Center for Equity, Diversity, and Inclusion, and was the founder of the UW School of Medicine's Center for

Cultural Proficiency in Medical Education. Dr. Acosta earned his bachelor's degree in biology from Loyola University and his medical degree from the UC, Irvine, School of Medicine. He completed his residency training at Community Hospital of Sonoma County in Santa Rosa, California, an affiliate of UC San Francisco School of Medicine, and a faculty development fellowship at the University of Washington Department of Family Medicine.

Victor J. Dzau, M.D., is the president of the National Academy of Medicine (NAM). He is chancellor emeritus and James B. Duke Professor of Medicine at Duke University and the past president and chief executive officer of the Duke University Health System. Previously, Dr. Dzau was the Hersey Professor of Theory and Practice of Medicine and Chairman of Medicine at Harvard Medical School's Brigham and Women's Hospital, as well as chairman of the Department of Medicine at Stanford University. Dr. Dzau has made a significant impact on medicine through his research in cardiovascular medicine and genetics, his pioneering of the discipline of vascular medicine, and his leadership in health care innovation. In his role as a leader in health care, Dr. Dzau has led efforts in health care innovation. His vision is for academic health sciences centers to lead the transformation of medicine through innovation, translation, and globalization. He has been a member of the Council of the Institute of Medicine (IOM), and the Advisory Committee to the Director of the NIH, as well as chair of the NIH Cardiovascular Disease Advisory Committee and the Association of Academic Health Centers, among other leadership positions. He is the recipient of many honors, including 10 honorary doctorates.

Howard Ross is a lifelong social justice advocate and the founder of Cook Ross. His books on identifying and addressing unconscious bias include *Everyday Bias: Identifying and Navigating Unconscious Judgments in Our Daily Lives*; *ReInventing Diversity: Transforming Organizational Community to Strengthen People, Purpose and Performance*; and *Our Search for Belonging: How the Need for Connection is Tearing Our Culture Apart.* He has led programs at Harvard University Medical School, Stanford University Medical School, Johns Hopkins University, the Wharton School of Business, Duke University, Washington University Medical School, and more than 20 other colleges and universities. From 2007 to 2008, he served as the Johnnetta B. Cole Professor of Diversity at Bennett College for Women, the first time a white man had served in such a position at an HBCU. He has been published by the *Harvard Business Review, Washington Post,*

New York Times, and other publications, and has been a regular guest on National Public Radio for more than 10 years. He has served on numerous nonprofit boards, including the Diversity Advisory Board of the Human Rights Campaign, Dignity and Respect Campaign, and National Women's Mentoring Network. He has received many awards for his contributions. Mr. Ross received his Bachelor of Arts in history and education from the University of Maryland and completed post-graduate work in leadership and management at Wheelock College.

Richard Rothstein is a distinguished fellow of the Economic Policy Institute and a senior fellow (emeritus) at the Thurgood Marshall Institute of the NAACP Legal Defense Fund. He is the author of *The Color of Law: A Forgotten History of How Our Government Segregated America.* He is also the author of many other articles and books on race and education, including *Class and Schools: Using Social, Economic and Educational Reform to Close the Black-White Achievement Gap* and *Grading Education: Getting Accountability Right.*

Vaughan Turekian, Ph.D., is the executive director of Policy and Global Affairs at the National Academies of Sciences, Engineering, and Medicine. From 2015 to 2017 he served as the science and technology adviser to the U.S. Secretary of State. In this capacity, he advised the Secretary of State and other senior State Department officials on international environment, science, emerging technology, and health matters affecting the foreign policy of the United States. Previously, he was chief international officer for the American Association for the Advancement of Science (AAAS) and the Director of AAAS's Center for Science Diplomacy. In this capacity, he worked to build bridges between nations based on shared scientific goals, placing special emphasis on regions where traditional political relationships are strained or do not exist. In addition, Dr. Turekian worked at the State Department as a special assistant and adviser to the Under Secretary for Global Affairs (2002-2006) on issues related to sustainable development, climate change, environment, energy, science, technology, and health. He is currently the co-chair of the 10-member group of experts advising the United Nations on science, technology, and innovation in support of the UN Sustainable Development Goals. He holds a B.S. in geology and geophysics and international studies from Yale University and an M.S. and Ph.D. from the University of Virginia, where he focused on the transport and chemistry of atmospheric aerosols in marine environments.

Harriet Washington is an award-winning medical writer and editor, and the author of the best-selling book, *Medical Apartheid: The Dark History of Medical Experimentation on Black Americans from Colonial Times to the Present*. In her work, she focuses mainly upon bioethics, history of medicine, African American health issues, and the intersection of medicine, ethics, and culture. The book won the National Book Critics Circle Nonfiction Award, a PEN award, 2007 Gustavus Myers Award, and Nonfiction Award of the Black Caucus of the American Library Association. Ms. Washington wrote *Medical Apartheid* while she was a Research Fellow in Ethics at Harvard Medical School. She has worked as a Page One editor for *USA Today* and as a science editor for metropolitan dailies and several national magazines. Her work has appeared in *Health, Emerge,* and *Psychology Today,* as well as such academic publications as the *Harvard Public Health Review, Harvard AIDS Review, Nature, Journal of the American Medical Association, American Journal of Public Health,* and *New England Journal of Medicine.* Her awards include the Congressional Black Caucus Beacon of Light Award, two awards from the National Association of Black Journalists, and a Unity Award from Emerge. She is the founding editor of *The Harvard Journal of Minority Public Health* and has presented her work at universities in the United States and abroad. Ms. Washington has taught at venues that include New School University, SUNY, the Rochester Institute of Technology, University of Rochester, Harvard School of Public Health, and Tuskegee University, and she has sat on the boards of many organizations. Ms. Washington has also worked as a laboratory technician, as a medical social worker, as the manager of a poison-control center/suicide hotline, and has performed as an oboist and as a classical-music announcer for WXXI-FM, a PBS affiliate in Rochester, New York.